T0207065

This book comprehensively describes all aspects of gravity flow, a physical process in the environment that is covered in many different disciplines, including meteorology, oceanography, the earth sciences and many industrial processes. The first edition of this book was very well received, and the new edition has been brought completely up to date.

No other book covers a similar range of information. The new edition contains a completely new chapter on gravity currents in satellite imagery, and many new sections, including glacier, purging in pipes, flow over porous media, and recent work on atmospheric bores and sea breezes. The physical processes of gravity currents are described with a variety of laboratory experiments, many from the author's own work.

Gravity Currents is a valuable supplementary textbook for undergraduates and a reference work for research workers in a range of disciplines, including fluid mechanics, meteorology, oceanography and earth sciences. The general reader will also find much of interest, since the physics of the flows involved is clearly described, without advanced mathematics, by numerous photographs and illustrations.

GRAVITY CURRENTS
In the Environment and the Laboratory

Second edition

Protection against the bore in the Qiantang River: an historic scene.

JOHN E SIMPSON University of Cambridge

Second edition

GRAVITY CURRENTS
IN THE ENVIRONMENT AND THE LABORATORY

CAMBRIDGE
UNIVERSITY PRESS

CAMBRIDGE UNIVERSITY PRESS
Cambridge, New York, Melbourne, Madrid, Cape Town,
Singapore, São Paulo, Delhi, Tokyo, Mexico City

Cambridge University Press
The Edinburgh Building, Cambridge CB2 8RU, UK

Published in the United States of America by
Cambridge University Press, New York

www.cambridge.org
Information on this title: www.cambridge.org/9780521664011

First published by Ellis Horwood 1987
Second edition first published by Cambridge University Press, 1997

First published 1999

A catalogue record for this publication is available from the British Library

Library of Congress Cataloguing in Publication data

Simpson, John E. 1915-
    Gravity currents in the environment and the laboratory /John E.
Simpson.- 2nded.
        p. cm.
Includes bibliographical references and index.
    ISBN 0 521 56109 4 (hbk)
    1. Gravity.   2. Fluid dynamics.   I. Title.
QB334.S56   1997
532'.051-dc20   96-14067 CIP

ISBN 978-0-521-56109-9 Hardback
ISBN 978-0-521-66401-1 Paperback

Contents

Foreword xi
Preface xii

1 The nature of gravity currents 1
1.1 Introduction 1
1.2 Dam break 3
1.3 Gravity currents 4
1.4 Bores 5
1.5 Internal bores 7
1.6 Solitary waves 9

2 Atmospheric gravity currents 11
2.1 Thunderstorm outflows 12
2.2 Frontal structure of a *haboob* 14
2.3 Atmospheric bores 21
2.4 Generation of bores 24
2.5 Development of a bore from a gravity current 25

3 Sea-breeze fronts 29
3.1 Formation and structure of sea-breeze fronts 30
3.2 Internal bores formed by the sea breeze 35
3.3 Sea breeze and pollution 37
3.4 Birds and insects at sea-breeze fronts 40

4 Gravity currents in satellite imagery 45
4.1 Use of satellites 45
4.2 Rope clouds 46
4.3 Ropes and cold fronts 49
4.4 Ropes in the sky 53
4.5 Hurricane ropes 56

5 Fronts and topography 59
5.1 The southerly buster 59
5.2 Other coastal trapped gravity currents 62
5.3 Katabatic winds and fronts 62

Contents

6 Environmental problems: atmosphere 66
6.1 Gravity current hazards to aircraft 66
6.2 Spread and dilution of a dense gas 69
6.3 Gravity currents of gases in mines 76
6.4 House ventilation flows 81

7 Gravity currents in rivers, lakes and the ocean 88
7.1 Ocean scale fronts 88
7.2 Fronts in estuaries 91
7.3 Fronts in fjords and lakes 94
7.4 Bores in the environment 95
7.5 Internal bores 99
7.6 Turbidity currents on the ocean bed 102

8 Industrial problems with gravity currents: oceanography 105
8.1 Power station effluents 105
8.2 Oil slicks 106

9 Avalanches 112
9.1 Types of avalanche 112
9.2 Snow avalanches 112
9.3 Rock avalanches (dry) 116
9.4 Debris and mud avalanches (wet) 118
9.5 Glaciers 122

10 Volcanic gravity currents 127
10.1 Basaltic lava streams 127
10.2 Pyroclastic plumes 128
10.3 Pyroclastic gravity currents 129
10.4 Mud flows or lahars 133
10.5 Eruption of Mount St Helens 135
10.6 The Lake Nyos gas disaster 136
10.7 Volcanic action on the ocean bed 137

11 The anatomy of a gravity current 140
11.1 The front 140
11.2 Inviscid-fluid theory 142
11.3 Effect of mixing on gravity current 144
11.4 Effect of friction 147
11.5 Head and tail ambient flows 152
11.6 Gravity currents meeting obstacles 155
11.7 Purging of fluid in a pipe 162

12 Spread of dense fluid 164

12.1 Lock exchange flows 164

12.2 Release of a fixed quantity of fluid in a rectangular channel 166

12.3 Radial collapse of a fixed quantity of fluid 172

12.4 Constant flux in a parallel channel 175

12.5 Constant flux, radial flow 177

12.6 Front moving along a free surface 178

12.7 Gravity currents on slopes 178

12.8 Gravity currents of high density ratio 180

12.9 Gravity currents over porous media 182

13 Ambient stratification 186

13.1 Two-fluid systems 186

13.2 Continuous stratification 197

14 Ambient turbulence 205

14.1 Turbulence experiments 205

15 Viscous gravity currents 213

15.1 Inertial and viscous regimes 214

15.2 The spreading of viscous gravity currents 215

15.3 Viscous flows on slopes 215

16 Suspension flows 218

16.1 Turbidity currents 219

16.2 Air suspension currents: fluidisation 221

16.3 Initiation of turbidity currents 224

16.4 Summary of suspension processes 225

17 Gravity currents on a rotating earth 227

17.1 Coriolis force 227

17.2 Rossby number 228

17.3 Spreading under gravity: no boundaries 229

17.4 Spread along a boundary 230

18 Numerical models of gravity currents 232

18.1 Marker-and-cell technique 232

18.2 Models of thunderstorm outflows 233

18.3 Analytical and numerical model 234

18.4 High-resolution models 236

Index 239

Foreword by Steve Thorpe
Professor of Oceanography, The University, Southampton, UK

The day of the gifted amateur scientist is past. No longer may we see his enthusiasm conveyed in a text which will both draw the attention of professional scientists and stimulate the interest of students. Books by learned academics, on the other hand, are sometimes so analytical and abstract that they fail to convey that element of surprise, even awe, which first arouses a spirit of enquiry and the formulation of questions, and leads to the development of experimental results and theoretical description.

Helped perhaps by his background, particularly a lifelong hobby, John Simpson has found a nice compromise. John was a school teacher when I first met him in the late 1960s and, for many years before, had been a very active glider pilot. He has flown over 1000 hours in 40 different types, and it was the effect of sea-breeze fronts on gliders which first interested him in that phenomenon. He 'composed' (his own words) a movie on the subject in 1967, and was awarded the Darton Prize of the Royal Meteorological Society for work published in the journal *Weather*. In 1971 he was appointed as a Research Assistant in the Department of Geophysics in the University of Reading (who must have counted themselves fortunate indeed!) and later, in 1976, he became a Research Associate in the Department of Applied Mathematics and Theoretical Physics at the University of Cambridge. There, in 1981, he gained his PhD, 45 years after the award at the same university of his BA. On a recent visit to China he surprised his hosts by asking to be allowed to see the bore on the Qiantang River (see sections 7.4 and figure 7.14), a rare request. His hobby had widened, and his enthusiasm shows.

In first perusing this book the reader will immediately be aware, as I was, of the number of clear and simple diagrams, the beautiful photographs which so vividly illustrate the text, and the extraordinary range of phenomena which belong to the genus 'Gravity Currents'. Many of them are of considerable practical importance in areas such as mining, the discharge of power station effluent or in the chemical industry, as well as in aircraft operations. Others are associated with some of the most impressive natural processes, for example sandstorms and volcanoes. John has himself contributed much to the explanation of these phenomena, especially by his careful laboratory studies. The book provides ample evidence of his academic, analytical approach to physical problems but is also a testimony to the author's delight in natural phenomena, conveying a spirit of wonder in the drama, magnificence and power of the geophysical fluids which are around us.

I welcome this book and hope that it may, in some readers at least, inspire a desire to see for themselves the phenomena it describes and to understand how they come about.

Preface

The need for this book was apparent several years ago when I prepared a short review on the subject of gravity currents. This had been limited to only 20 pages and as I looked into wider environmental aspects I could see the value of a more comprehensive review, of at least ten times the length, on the manifestations of gravity currents and the closely related internal bores and solitary waves. Workers in many scientific disciplines, who are not experienced in fluid mechanics, could benefit from an account of what is already known about the properties of gravity currents and would be able to see how much is applicable to their own specialist subject. As the work developed I could see that the material described, with the many photographs and diagrams I was able to include, would also be of interest to the general reader.

The first part of the book, from Chapter 1 to Chapter 10, deals with the nature of buoyancy-driven flows in the atmosphere and oceans and on the earth's surface. These flows in the atmosphere are at scales from a cold gravity current through an open door of a house to vast squall lines of cold dense air which are hazardous to aircraft. In the ocean the study ranges from the Gulf Stream to river fronts in estuaries and fjords. The formation of these fronts is of biological importance; other important gravity currents are heated effluents from power stations. The section on earth sciences deals with snow avalanches and volcanic gravity currents.

Industrial problems in which gravity currents play an important part are described; a topical example is the spread of a cloud of dense gas which may be poisonous or explosive. The study of gravity currents of gases in mines has a long history.

The second part of the book, Chapter 11 to Chapter 17, deals with 'the anatomy of a gravity current', and examines the many factors which affect their behaviour, including, as well as the nature of the head of the current, the influence of stratification and turbulence in the surroundings. The topics are dealt with by numerous laboratory experiments, and only simple mathematics is included. The final chapter mentions briefly the rapidly growing subject of numerical models of gravity currents.

References are collected together at the end of each chapter to avoid interrupting the text.

Many scientists kindly gave permission for use of striking photographs of phenomena in the environment; most of the laboratory photographs are my own.

I am grateful for the help of Professors R.S. Scorer and S.A. Thorpe and I am indebted to many other friends, especially the following, each of whom commented on the chapter dealing with his speciality: Drs Tim Davies, Herbert Huppert, Jim McQuaid, Alf Mercer, Roger Smith, Steve Sparkes and Alan Thorpe. Dr Chris Bertram read the whole manuscript and suggested many improvements in the presentation. Members of the Department of Applied Mathematics and Theoretical Physics at Cambridge have helped me in many ways; Drs Paul Linden and Jim McDonnel made many helpful comments. My thanks go to Margaret Downing for her able preparation of the diagrams.

Preface to the second edition

In the new edition I have tried to bring the book up to date, and have added some completely new sections.

The new material includes a chapter on gravity currents in satellite imagery, and sections on glaciers, purging in pipes and flow over porous media. Recent field work on the generation of atmospheric bores and on gravity current frontogenesis in the sea breeze has been included. I have dealt with the progress of warnings to aircraft of wind-shear hazards, and of buoyancy driven ventilation flows in buildings. Important new laboratory work on gravity currents with large density differences is included.

Again I am indebted to members of the Department of Applied Mathematics and Theoretical Physics at Cambridge for many helpful discussions and especially to Drs John Bush and Colin Wilson who have suggested many helpful alterations to the text.

I am grateful to Dr G. Kappenberger and J. Hacker for the use of their photographs on the cover of the book.

1   The nature of gravity currents

1.1 Introduction

Gravity currents, sometimes called density currents or buoyancy currents, occur in both natural and man-made situations. These currents are primarily horizontal flows and may be generated by a density difference of only a few per cent.

An important part is played in many different scientific disciplines by gravity currents. In the atmosphere, for example, most of the severe squalls associated with thunderstorms are caused by the arrival of an enormous gravity current of cold dense air. One such advancing atmospheric gravity current in the Sudan is shown in figure 1.1. In this case the dense air, which is moving from right to left in the picture, is clearly outlined by sand and dust which have been raised from the ground by the strong turbulent wind. The dust cloud is about 1000 metres high and the front is advancing at about 25 m s^{-1}; some idea of the scale can be gained from the houses which can just be made out in the distance.

Figure 1.1 The front of a gravity current of cold air in the atmosphere, made visible by suspended sand and dust.

Knowledge of the properties of these gravity currents is obviously important for aircraft safety. The fronts produce large changes in horizontal wind and areas of intense turbulence. As they are not always so clearly marked by dust as the example in the photograph, it is possible to fly into them without any warning. Encounters of this kind have been responsible for serious accidents, both at take-off and at landing.

Another, less intense, manifestation of atmosphere gravity currents appears in the sea-breeze front. These fronts form near the coast, and many of them propagate up to 200 km inland. They have important effects on the transport of airborne pollution, and also on the distribution of insect pests.

Avalanches of airborne snow, which are a severe hazard in the mountains, are gravity currents in which the density difference is supplied by the suspension of snow particles. For many years attempts have been made to reduce the damage caused by avalanches and there are research establishments devoted solely to the investigation of this special type of gravity current.

An industrial problem which has received much attention recently is the accidental release of a dense gas, which may be poisonous or explosive. Serious accidents have occurred in the resulting spread, which usually starts as a gravity current. Much experimental and theoretical work has been carried out on this problem, leading to possible methods of controlling such escapes.

Even in the home, problems with gravity currents are common. If the door of a warm house is held open for a few seconds on a cold day it is easy to detect the gravity current of cold air flowing along the ground into the house.

This open door experiment is recommended to the reader, who may care to use soap bubbles or puffs of smoke to detect the sudden onset of the gravity current of dense cold air after the door has been opened. This topic is dealt with in more detail in Chapter 14.

In the ocean, large volumes of warm or fresh water, less dense than the neighbouring salty water, flow as gravity currents along the surface. Gravity currents in the ocean are not as obvious to the casual observer as some atmospheric gravity currents, but lines of foam and debris on the surface may point to their presence. These lines are caused by the convergence of the flows there, and are well known to fishermen, since these currents have important effects on the distribution of fish.

Fresh-water gravity currents often flow along the surface in estuaries and fjords, above the more dense sea water. Figure 1.2 shows an echo-sounding of such a surface flow, made in the Fraser River in Canada. This shows the cross-section of a gravity current of fresh water advancing from the right, above the denser salt water from the sea. The leading edge of this current has a 'head' which is deeper than the following flow, a feature which is seen in most gravity currents.

The oil slick is an example of a man-made environmental problem. An oil spillage from a ship forms a non-mixing gravity current of less dense fluid on the sea surface. It is important to understand the development of this flow and find possible methods for both its containment and its dispersal.

1.2 Dam break

To understand the physics of a gravity current it will help to consider what happens when the wall of a dam breaks and releases the water behind it.

A mass of suddenly released water will start to collapse and flow horizontally. The main force acting on the water in such a flow is due to gravity, and acts vertically downwards. This results in a downward motion of the water which can only occur if the water spreads horizontally. So the potential energy of the water due to its height is continuously converted into the kinetic energy of the horizontal motion.

If the flow spreads mainly in one direction, for example along the bottom of a valley, a rough idea of the velocity, U, in that direction can be obtained by equating the values of the potential energy loss and the kinetic energy gain, i.e.

$$(mU^2)/2 = mgH/2$$

or

$$U = \sqrt{(gH)}$$

where m is the mass, $H/2$ the mean height of the centre of gravity and g the acceleration due to gravity. If for example the water was originally 20 m deep, the velocity would be about 14 m s^{-1}, or roughly 30 mph.

Viscous forces in the fluid can also have important effects. Viscosity may be likened to friction, in that a viscous fluid exerts retarding forces on those parts of itself which are trying to move with greater velocity than the rest, just as retarding forces due to friction occur between two solid surfaces in relative motion. The lower layers of the water in the dam-break flow are retarded by the

Figure 1.2 An echo-sounding made in the Fraser River in Canada. The front of a gravity current of fresh water is moving above sea water. (Courtesy of David Farmer.)

ground, and have a considerable effect on the form of the leading edge of the fluid.

1.3 Gravity currents

The water in a dam-break flow is submerged in the atmosphere, but this has only a very small effect on its behaviour. If the air is replaced by a fluid which is only a few per cent less dense than the collapsing fluid, then the flow will be different.

If in this 'dam-break analogy' flow the density difference between the two fluids were only 1%, the effective driving force would then have been reduced to only 1% of normal. Unless the coefficient of viscosity is large, or the scale is very small, the main controlling forces will be gravitational and inertial, due to the displacement of the fluid around the advancing current. Due to the net gravitational acceleration of the collapsing fluid being now only $g/100$, the previous dam-break flow will be replaced by one appearing to move 'in slow motion'.

A typical *gravity current* of dense fluid is now moving forwards into a slightly less dense fluid. In this case its rate of advance U can be approximated by

$$U = (gH/100)^{\frac{1}{2}}$$

or, in general, if ρ is the density of the less dense fluid and $\Delta\rho$ is the density difference,

$$U = \left(\frac{\Delta\rho}{\rho}gH\right)^{\frac{1}{2}}$$

The term $g\,(\Delta\rho/\rho)$ will usually be denoted by the symbol g', and called 'reduced gravity'.

The fluid in a gravity current may be chemically different from the surroundings and have a different molecular weight, but often the difference in specific weight that provides the driving force is due to dissolved material or to temperature difference. The large-scale gravity current in the atmosphere shown in Figure 1.1 was caused by temperature differences. If the temperature was about 12°C, this would give a density difference of about 4%. The value of g' will be 0.39 m s^{-2}. With a current height of 1000 m, we would expect the rate of advance to be about $(g'H)^{\frac{1}{2}}$ which is just under 20 m s^{-1}.

1.3.1 Suspension flows; turbidity currents

One way in which the overall density of a fluid can be increased is by the suspension of many small dense particles within it. Such suspension currents may be formed in various ways. One of the most important processes is the raising of material from the ground and its suspension by the turbulence within a gravity current. This suspended material increases the density and hence the speed and turbulence within the current. The process can thus become 'self stoking' in a current on a slope, further increasing the strength of the current. An example of such a suspension current is shown in Figure 1.3. This illustrates a suspension current of kaolin in water advancing through water towards the camera in the laboratory. This photograph shows very clearly the complicated shifting instability patterns manifested by a gravity current advancing along a plane surface.

Self-stoking gravity currents containing suspended matter also occur in the ocean. They start on slopes near the coast as mud-slides which increase in intensity until a suspension current is formed. These *turbidity currents* may become large enough to travel at speeds of over 30 m s^{-1}. They can gouge out vast channels in the sea bed and their progress has been followed by monitoring the successive breaking of submarine telephone cables, showing that they can travel for hundreds of kilometres.

1.4 Bores

A related phenomenon is the *bore*, which is also concerned with mass transport and has many features in common with the gravity currents already described.

The best-known type of bore is a tidal disturbance which moves upstream in some rivers and may be very violent at spring tides. It is an example of a hydraulic jump in which there is a sudden increase of the water depth associated with a change in the flow rate.

If the increase of depth at the front of a bore is less than about a third of the undisturbed water depth, a series of smooth waves appears at the leading edge and the bore is called 'undular'. For larger steps the tidal bore is turbulent and it advances as a wall of tumbling breakers; the structure is similar to that of breakers advancing towards the sea-shore.

Bores appear in rivers where the tidal range is large and the form of the estuary is suitable. Conditions are favourable in several rivers in England: on the River Severn the bore has been ridden by experienced surfers and journeys of

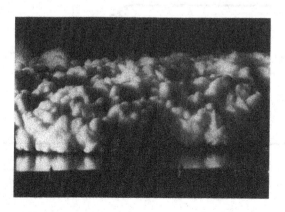

Figure 1.3 A suspension current in the laboratory, advancing towards the observer. (Courtesy of J.H.R. Allen.)

two or three miles have been achieved. However, since they are produced by tides, surfers who miss their moment may have to wait over 12 hours for the next breaker! Figure 1.4 shows a rider on a surf board using one of the waves at the front of the Severn bore. The bore shown here is an undular one, but breaking waves are forming in the shallower water near the banks of the river.

The conditions ahead of a hydraulic jump or a bore can be related to those behind it in a simple mathematical treatment.

In figure 1.5 the jump is shown brought to rest in a moving frame of reference; h_1, U_1 and h_0, U_0 are the height and velocity on the two sides of the jump.

Figure 1.4 A surfer on the advancing bore on the River Severn. (Courtesy of the Severn–Trent Water Authority.)

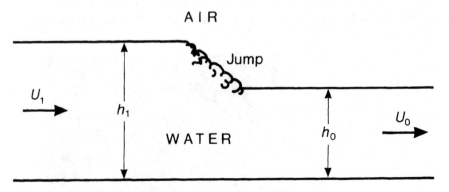

Figure 1.5 The flow through a hydraulic jump. The axes have been taken to bring the jump to rest, and h_1, U_1 and h_0, U_0 are the height and the velocity on the two sides of the jump.

The volume flux Q per unit width, passing in unit time, is, since mass is conserved,

$$Q = U_0 h_0 = U_1 h_1$$

A change in momentum across the jump is caused by the pressure difference. If density is ρ, since mean pressures at both sections are $g\rho h_0/2$ and $g\rho h_1/2$, the equation of momentum is

$$Q(U_1 - U_0) = g(h_1^2 - h_0^2)/2$$

hence

$$Q^2 = g h_0 h_1 (h_1 + h_0)/2$$

The energy, however, does not balance and the loss of energy per unit time is

$$\rho(U_0^2 - U_1^2)/2 + g\rho(h_0 - h_1)$$

which can be shown to be

$$g\rho(h_1 - h_0)^3/4h_1 h_0$$

This loss of energy, which must occur at a bore, is mostly effected in the undular case by the waves, each of which carries energy as it moves away from the front. The more intense bores cannot carry away enough energy by this method and the energy excess is dissipated by turbulence in the tumbling breakers at the leading edge.

1.5 Internal bores

The previous section considered bores at the free surface of a water flow. A somewhat similar class of internal bores can be formed at an interface between two fluids, one lying on top of another which is perhaps only a few per cent denser. Compared with surface bores, these internal bores appear to move 'in slow motion', since the buoyancy forces are very much reduced.

Internal bores have been described theoretically and investigated in laboratory experiments. Figure 1.6 shows such an experimental arrangement in which an obstacle is towed along the bottom of a tank containing a two-fluid system. The obstacle is moving to the right and displacing an undular internal bore which steadily moves along the interface ahead of it. If the moving obstacle is replaced by an advancing gravity current of dense fluid a similar effect can be produced.

Such experiments have been very helpful in understanding the physics of undular bores and will be described in more detail in Chapter 13.

During the last few years, internal bores have provided an explanation for an increasing number of phenomena in the environment, both in the ocean and in the atmosphere. They may for example be formed in the ocean by tidal effects on fresh-water layers near the coast. In the atmosphere they are formed in dense stable layers by advancing flows of cold dense air from thunderstorms, and they are also associated with sea-breeze fronts.

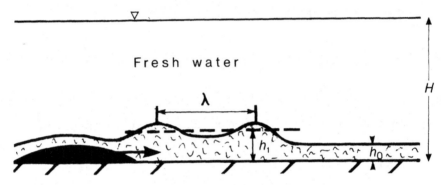

Figure 1.6 The production of an internal bore by a moving obstacle (black) in an experimental laboratory tank, in which a layer of fresh water (clear) lies above salt water (shaded). The bore is moving to the right along the interface between the two fluids, faster than the moving obstacle.

Figure 1.7 Clouds marking an undular internal bore in the atmosphere, the 'Morning Glory' in North Australia. (Courtesy of Roger Smith.)

When the ratio of the height h_1 behind the jump (see figure 1.6) to that in front of it, h_0, is less than about 2, then the internal bore is undular. In much deeper internal bores the leading edge is turbulent and appears very similar to the front of a gravity current. The photograph in figure 1.7 shows the clouds forming at an atmospheric undular bore in Northern Australia. This phenomenon appears in the early morning and is marked by a spectacular roll of cloud; its striking appearance has led to its name, the 'Morning Glory'.

1.6 Solitary waves

'Gravity currents' and 'gravity waves' are sometimes confused and the distinction between them is not always apparent. Here the name 'gravity current' will be applied to a phenomenon in which there is a clear transfer of mass (usually horizontal). In gravity waves there is little transfer of mass and the main transport is that of energy.

An undular bore, as has been noted, consists of an increase in depth of a fluid advancing with a series of waves on its surface. Closely related is the 'solitary wave' which is another shallow-water phenomenon, i.e. a disturbance which is high compared with the undisturbed depth. A solitary wave is not a

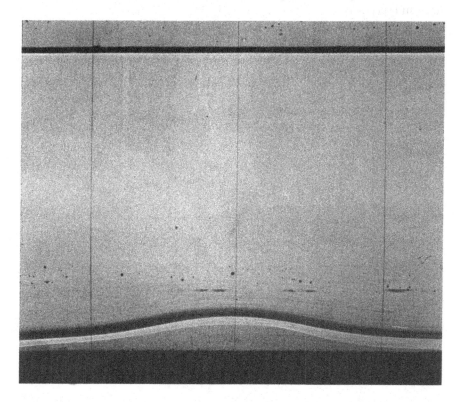

Figure 1.8 Internal solitary wave moving along an interface between two fluids in a laboratory tank.

periodic wave but consists of a single symmetrical hump which propagates at uniform velocity without change of form.

Scott Russell in the 1840s investigated solitary waves on the surface of a canal. On horseback he was able to follow examples of such waves for several kilometres and he showed that, with length, depth and amplitude properly matched, a solitary wave can propagate virtually unchanged, except for small effects due to bottom friction which reduce the size of the wave.

Internal solitary waves can exist at an interface between two fluids of different density. Examples occur in the atmosphere, where solitary waves have been observed on stable layers, moving steadily away from the distant disturbances which generated them. What is observed on the ground is somewhat similar to the arrival of a bore, with a line of cloud and a gust of wind, but in this case with only a temporary increase of surface pressure.

In the laboratory these internal solitary waves are very easy to create. Figure 1.8 shows an example of an internal solitary wave formed at a layer which had been laid down by a gravity current. When the gravity current reached the end of the tank some of the dense fluid ran up the wall and then descended as a mass which generated the solitary wave shown, moving from right to left. When this disturbance reached the other end of the tank, three metres distant, it was reflected and returned with nearly the same shape and speed.

2   Atmospheric gravity currents

Gravity current fronts and internal bores have many manifestations in the atmosphere, some at a very large scale. A number of case studies of atmospheric observations will be examined to see how far they can be explained in terms of the physical processes described in Chapter 1.

Experiments with dense saline flows in water tanks have elucidated features of dense atmospheric flows. This is because many of the results of buoyancy in a compressible atmosphere can be deduced from an incompressible fluid if potential temperature is substituted for density (with the difference in sign noted). The potential temperature of the air at any point is the temperature this air would attain if its pressure were changed to 1000 mb (standardised ground level), with no gain or loss of heat.

A zone of constant potential temperature in the compressible atmosphere can be modelled in the laboratory by a salt solution of constant density. A tank of salt solution with its concentration and density decreasing with height is stable and corresponds to a stable atmospheric zone with potential temperature increasing with height.

The *dynamical similarity* between the gravity currents seen in the laboratory and those seen in thunderstorm outflows has been established in many case studies (Simpson & Britter, 1980. Idso *et al.*, 1982). The requirements for formal similarity are (A) geometric similarity between the model and the large-scale flow, and (B) the equality of the relevant dimensionless numbers. In this context the two most important dimensionless numbers are the *Reynolds number*, Re, and the *Froude number*, Fr.

The Reynolds number is a dimensionless ratio which gives a criterion for the critical behaviour of a flow, depending on its speed, depth and coefficient of viscosity. The Reynolds number of a flow at velocity U, depth h and kinematic viscosity ν is given by Uh/ν. Reynolds showed, by using a slender trail of dye through the centre of a pipe, that the flow would change from a laminar to a turbulent one for Reynolds numbers greater than about 600.

As will be seen in the descriptions of gravity current experiments, when the Reynolds number is greater than about 1000 the flow patterns are independent of its value. As the Reynolds number in most thunderstorm outflows is of the order of 10^8, this ratio is unimportant in determining the nature of the flow.

The most important dimensionless number related to the flow of atmospheric gravity currents is the internal Froude number, the ratio of inertial forces to buoyancy forces, usually written in the form $Fr = U/(g'h)^{\frac{1}{2}}$. The velocity of the current is U, g' is reduced gravity and h is the depth where the density discontinuity occurs. The Froude number of cold outflows when they have been measured is not very different from 1, in agreement with the smaller-scale laboratory results.

Laboratory experiments have been able to include a number of conditions of the ambient fluid, such as uniform velocity with a simple shear profile, and density profiles with a step and with uniform variation. It has not been possible to model all the complicated atmospheric configurations of wind and density profiles, and other sources of information may be able to supplement laboratory results. One important source is the numerical model, as described in Chapter 18.

2.1 Thunderstorm outflows

Thunderstorms are generated by warm moist air rising in unstable conditions. Eventually, this rising air reaches the boundary formed by a very stable layer called the tropopause. Here it spreads out, forming the familiar 'anvil' cloud. During the early stages of anvil development, the ice-crystal cloud sometimes gives a good indication of the form of the flow patterns. An example of this is shown in figure 2.1, which was taken from an aeroplane just beneath a

Figure 2.1 Clouds at an active thunderstorm, showing developing anvil cloud spreading to the right above the storm. At 30 000 feet above Turin, on 12 August 1972. (Courtesy of Colin Street.)

developing anvil cloud at about 30 000 ft above Turin in Italy. This anvil cloud can be seen to have the form of a gravity current moving towards the right beneath a rigid boundary. It has the typical contour of the front, but the bulges visible beneath it are probably due to fall-out of ice crystals which often form 'mamma' cloud.

Rain and hail, falling in another part of the thunderstorm, produce a downdraught of cold air that descends towards the ground. When the cloud-base is high, the rain falls an appreciable distance through this non-saturated zone and the cooling due to evaporation may be very great. The cold column of air, which may be several per cent denser than the surrounding air, reaches the ground and spreads out horizontally, forming a gravity current.

A schematic diagram of the structure of a thunderstorm cell is shown in figure 2.2. The downdraught which spreads out away from the storm is shown here as falling behind the area of rising currents, but in fact it moves perpendicular to the plane of the paper – so this really needs to be demonstrated in three dimensions. The downdraught section of a thunderstorm plays an important part in the structure and regeneration of severe storms, since the dense cold air pushes up the warmer and lighter air it meets. Thus, it reinforces the warm updraft that creates new rain clouds in a system of thunderstorm cells.

2.1.1 Outflow observations from the ground
When a thunderstorm outflow advances over sandy and dusty land in an arid country, the strong turbulent wind following the gust front raises dust from the ground. One of the most awe-inspiring sights in nature then appears, a vast wall of dust 1000 metres high advancing across the country. The appearance of such

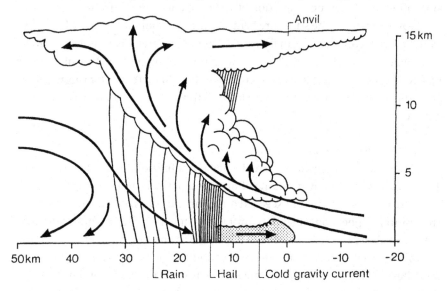

Figure 2.2 Schematic diagram of a thunderstorm cell. Flows shown are relative to the storm, which is travelling to the right.

a dust-laden cold outflow was shown in Figure 1.1. These fronts of cold dense air may move at 20 m s^{-1}, with a wind behind the first squall gusting up to twice this speed.

These dust storms occur regularly in Arizona, USA, where their approach to the town of Phoenix has been described by Idso *et al.* (1982). They appear in India, where they have been given the name of 'Andhi'. They are also common in many parts of Africa, especially in the Sudan, where they are named 'haboob', which is merely the Arabic name for strong wind (Sutton, 1951).

2.2 Frontal structure of a *haboob*

Haboobs are particularly prevalent near Khartoum, where they are most common in June. A *haboob* approaching the airfield at Khartoum on the evening of 17th June 1969 was studied in detail and a time-lapse movie was made to observe changes in the frontal structure. Measurements were taken of the wind and temperature changes, and the dust content was also recorded (Lawson, 1971).

The forward movement of the front as seen from the ground was in many ways similar to that described in laboratory experiments with gravity currents. A push of cold air from within the body of the *haboob* caused the formation of a bulge, or lobe, which grew forward faster than the average speed of the front. The lobe expanded until its speed relative to the front began to decrease and then irregularities appeared, with a fresh surge developing from one side. So new lobes were continually appearing out of the dying stages of a previous one. As in laboratory experiments, the convergent movement at the edges of two adjacent lobes often involved much ingestion of the fluid ahead of the front.

Some details of the structure behind the front could be deduced from the temperature records, shown in figure 2.3. These measurements were made at roof height every 1.6 seconds. They show an initial temperature drop of between 1.5 and 2°C, occurring over a distance of less than 50 metres. After this rapid drop, the mean level decreased only slowly, falling another half degree in ten minutes.

2.2.1 Dust content

Samples of the airborne dust were collected in visibilities ranging from 100 m to 3 km by passing a known volume of air through filters and weighing them. The increase in weight never exceeded 0.1 g, the minimum detectable difference. The resultant upper concentration limit was 40 mg m^{-3}, equivalent to a density difference about 1/300 of that due to the temperature drop between the warm and cold air, so it appeared that the dust content could be neglected as a contributing factor to *haboob* dynamics. Similar results for dust content have obtained in measurements made for summer dust storms in Russia, in the Mangyshlak Peninsula (Kharitanova, 1969).

2.2.2 Pressure changes at the surface

After the particularly dry summer in SE Australia of 1983, several dust squalls were observed in the Melbourne area. Figure 2.4 shows the front of one of these 'haboobs' crossing the city, where the buildings help to give an idea of the scale of the phenomenon.

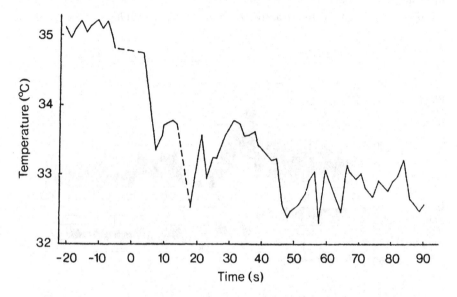

Figure 2.3 Temperature record obtained during passage of a *haboob* front at Khartoum, 17 June 1969. The time is relative to the arrival of the front. (After Lawson, 1971.)

Figure 2.4 A *haboob* crossing the city of Melbourne, 2 February 1983. (Courtesy of Melbourne Weather Bureau.)

A particularly clear record of a dust squall was obtained by Garratt (1984) as it passed Aspendale, Victoria; this is shown in figure 2.5. On the 8th day of February at 1500 h the wind shifted from about 5 knots North to strong southerly. The front of the dust storm advanced at 20 m s^{-1} and the speed of the initial gust was usually greater than this. The temperature fell from 35 to 28°C at the squall, and the pressure rose suddenly. Assuming that the pressure change was hydrostatic and due to the arrival of cold air of a height h, then the depth of

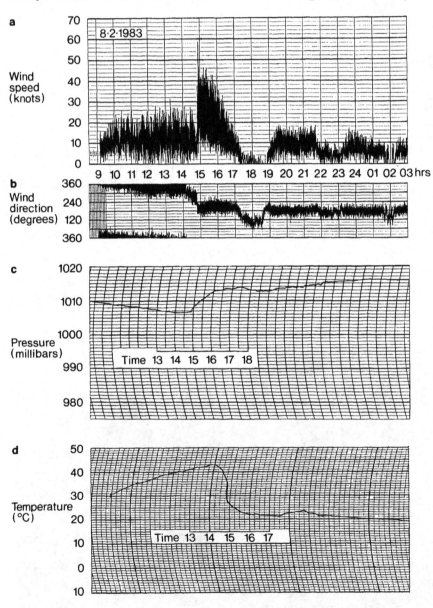

Figure 2.5 Surface records of passage of a *haboob* at Aspendale, Victoria, Australia, 8 February 1983. (a) Wind speed; (b) wind direction; (c) atmospheric pressure; (d) temperature. (Courtesy of J.R. Garratt.)

the dense air can be calculated. This gave an initial head height of 320 m, gradually increasing up to 1000 m within one hour.

If we interpret the flow as a gravity current, the velocity of the front is then given by

$$U = k(g\Delta T\, h/T)^{\frac{1}{2}}$$

If we take $k = 1$, this gives a velocity of $22\ \text{m s}^{-1}$, very close to the observed value of $20\ \text{m s}^{-1}$.

The detection of these pressure changes at the front of a thunderstorm outflow, using an array of pressure sensors on the ground, has been used to trace its progress across country.

2.2.3 Measurements from instrumented towers

The depths of cold outflows from thunderstorms vary from only a few hundred metres to well over a kilometre and much useful information of their structure has been obtained from measurements from instrumented towers (Hall *et al.*, 1976). Television towers have often been used but some research organisations have their own fully equipped instrumented towers. From one such multi-level tower 461 m high in central Oklahoma (Goff, 1976b) measurements of wind and temperature have been taken during the passage of many gust fronts and profiles of airflow and of temperature have been constructed.

A model of a typical outflow has been built up from these measurements, and is shown in figure 2.6. The raised head and the internal flow show a similar structure to that seen in the laboratory gravity currents we will describe in later chapters. In 17 cases examined, the height of the foremost point, or nose, was around 100 m. It was found that low-level stratification affected the slope of the gust front, and that the outflow could sometimes override a dense layer near the ground, as in the intrusive flows described in Chapter 13.

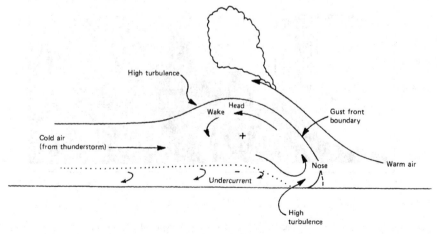

Figure 2.6 Schematic diagram of thunderstorm outflow, built up from tower measurements. (After Goff, 1976b.)

In several examples of gust fronts there were secondary or even multiple surges, whose behaviour did not fit into the simple gravity current picture. Some of these appear to be related to the generation of an atmospheric bore by the gravity current.

2.2.4 Remote sensing: sonar

Although tower data provide much useful information on cold outflows, they suffer from a number of limitations. By their nature, towers can only reach a few hundred metres and outflows may be three or four times the height sampled. There is also a certain amount of luck in having a non-portable tower in the right place to catch an outflow.

Although echo sounding in the ocean dates from the 1920s, it was not until 1968 that effective echo sounding was achieved in the atmosphere (McAllister, 1968). Improved antennae soon permitted operation at about 1000 Hz, giving a range of up to 1.5 km.

Each pulse of 'acoustic radar' or *sonar* produces on the facsimile recorder a vertical trace whose darkness varies with the received signal from a particular height in the atmosphere. These traces are closely spaced, so a trace is obtained of the evolution of atmospheric structure with time.

The acoustic scattering which can be detected is produced by small-scale temperature fluctuations associated with turbulence in regions of temperature and wind gradients, so we would expect to see strong echoes from a gravity current front. Ground-based acoustic sounding has been found to be effective in remote sensing of cold outflows and figure 2.7 shows an acoustic sounder record of a storm outflow passing Haswell, Colorado (McAllister, 1968). The vertical noise-lines in the upper part of the record at 2320 are caused by wind noise associated with the passage of the front. Otherwise, the echoes show clearly the boundaries of the turbulent outflow, and also give some information about the internal structure of the gravity current.

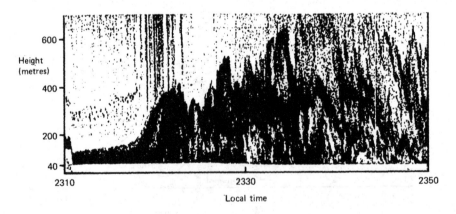

Figure 2.7 Acoustic sounder record of thunderstorm outflow passing Haswell, Colorado. (Reproduced with permission from L.G. McAllister, 1968, copyright Pergamon Books Limited.)

2.2.5 Remote sensing: radar

Radar has been used widely in building a model of the dynamics of the front of a cold outflow. Doppler radar, in which the change in frequency from moving targets is used, enables wind speeds to be measured at large heights and distances. The results from many tower measurements combined with radar information confirm the expectation that most atmospheric cold outflows are gravity currents with similar characteristics to those investigated in the laboratory. A narrow cold frontal band, accompanied by strong winds, tornadoes and pressure jumps, was examined by triple doppler radar as it passed through the Central Valley of California (Carbone, 1982). The isometric plot shown in figure 2.8 reveals a 'classic gust front' surface.

Using the Froude number relationship for a gravity current, the expected velocity for this front has been calculated. A more exact form, including the pressure term, gives

$$U = \{ \text{Fr} \, g\Delta z \, [(P_2/P_1)T_1 - T_2]/T_2 \}^{\frac{1}{2}}$$

where P_1, T_1 and P_2, T_2 are pressure and temperature before and after the passage of the front.

The pressure increase from numerous microbarographs recorded a surface pressure increase of 3 mb, giving results of $U = 21.4$ m s^{-1} and 18.3 m s^{-1} for values of Fr $= \sqrt{2}$ and 1.1, respectively. The observed speed of the gust front was 21.7 m s^{-1}, suggesting that the gravity current hypothesis is a likely explanation of the storm motion.

The lobe and cleft structure at a *haboob* has already been described – it appears similar to those seen in laboratory experiments. In the overall view of figure 2.8 some lobe structure is also apparent at the leading edge. Measurements of precipitation behind mesoscale cold fronts show regions of precipitation core divided by gap regions. It has been suggested that these oriented, ellipsoidal precipitation cores and gap regions are similar to the bulges and clefts and could be due to an instability produced by the strong horizontal shear of the wind across the front.

The structure of most of the atmospheric fronts described has been based on an almost instantaneous 'snapshot' of a front, or built up from a tower as the front passed by, assuming the structure did not change during the time of

Figure 2.8 Perspective view of gust front, measured by doppler radar.
(Courtesy of R.E. Carbone.)

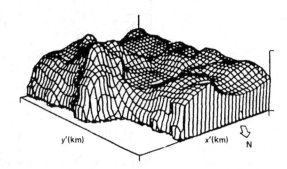

y'(km)   x'(km)

N

passage. Particularly in the early stages of development such observations may not give a true picture.

In a doppler radar study, allowing determination of the entire vertical structure, the life cycle of the gust front has been divided into stages (Wakimoto, 1982). A conceptual model of the early stages of the evolution of a gust front as measured in the atmosphere is shown in figure 2.9. The rain which gave the radar echoes showed first a short period of horizontal spread (Stage 1 in figure 2.9). In Stage 2 a roll-up commenced, and by Stage 3 a marked 'precipitation roll' had appeared, and moved forward, almost cut off from the following dense flow (Stage 4). The form of this flow appears to be identical with the behaviour of the current in the laboratory experiment pictured in figure 12.8 and described in Chapter 12. Here, it is sufficient to note that a rapidly changing structure occurs in the early stages of a developing divergent flow.

2.2.6 Remote sensing: lidar

Laser radar, usually called lidar, was first used to observe aerosol distribution in the atmosphere in 1963. Lidar receives back-scattered light from airborne particulates, and has the advantage that it does not require any turbulent fluctuations in atmospheric temperature or humidity, as usually employed in sonar and radar observations.

Lidar has been used in the atmosphere (Shimizu, *et al.*, 1985) employing the plan position indication (PPI) and range height indication (RHI) modes. Both the structure of the convective mixed layer and fronts of gravity currents have been studied. The RHI display of one of the latter will be described in more detail in Chapter 3.

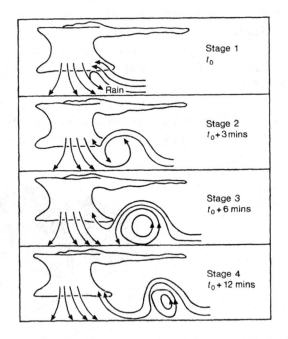

Figure 2.9 A model of the early stages of the evolution of a gust front, from doppler radar measurements. (After R.M. Wakimoto, 1982.)

Stage 1
t_0

Rain

Stage 2
$t_0 + 3$ mins

Stage 3
$t_0 + 6$ mins

Stage 4
$t_0 + 12$ mins

2.3 Atmospheric bores

A variety of atmospheric observations, especially during the last ten years, has shown what seem to be internal bores propagating on stable layers such as temperature inversions. An abrupt increase in ground level pressure (several millibars in a few minutes) is followed by a sustained period of high pressure, often containing wavelike oscillations. The increase in ground level pressure is often accompanied by an increase in ground level temperature and a shift of wind direction to that in which the disturbance is moving. There may be distinctive cloud formations, sometimes a roll cloud, perhaps followed by a series of rolls. These disturbances can travel several hundred kilometres from their point of generation. A study of undular pressure lines has been made in the arid region of northern Australia, and their development has been traced over long distances (Christie *et al.*, 1979).

Some early observations of these pressure jump lines were made in the midwest United States by Tepper (1950), and it was speculated that these disturbances were generated by an impulsive motion of a cold front into an existing nocturnal inversion, producing a propagating bore or solitary wave. As will be seen in Chapter 13, it is possible for the steady advance of a front of cold fluid to generate an internal bore in a stable layer over a wide range of depths and density differences.

Similar case studies made in North America by Schreffler & Binkowski (1981) show that the cold outflows from thunderstorms may act as a source of pressure jump lines, or atmospheric internal bores.

2.3.1 The Morning Glory

In the southern hemisphere, similar atmospheric bores have been observed (Clarke *et al.*, 1981; Smith *et al.*, 1982). The most spectacular is the so-called 'Morning Glory' which occurs in northern Australia near the southern coast of the Gulf of Carpentaria. This usually appears as a series of roll clouds, accompanied by a wind squall and a sharp rise in surface pressure. Figure 2.10 shows its arrival from the south soon after sunrise. Several expeditions have been made to investigate the Morning Glory, and the resultant photographs which display the various forms of an internal bore will be described in Chapter 13.

In the form 'b' seen in the laboratory experiments in Chapter 13 (see figure 13.2), rolls are very clear, but turbulence appears in the rear face of the first roll and extends to those which follow. A cloud formation displaying very similar behaviour is shown in figure 2.11.

As well as investigations by aeroplane, measurements of the flow patterns have been made by pilot balloon ascents, and a typical example is shown in figure 2.12. This shows the streamlines up to a height of 1500 m, deduced from seven balloon ascents across a distance of 69 km. The height of the top of the stable layer shows that the strength (or height ratio) of the bore is about 2 and that the wavelength of this undular bore is about 10 km.

Figure 2.10 Roll cloud marking the arrival of the Morning Glory at Burketown, northern Australia, at 0630 local time on 12 October 1980. (*Courtesy of Roger Smith.*)

Figure 2.11 The Morning Glory on 11 October 1981. Aerial view, showing a series of roll clouds NE of Burketown, Australia, looking east. (*Courtesy of Roger Smith.*)

The more intense Morning Glories have a strength much greater than in this example, and figure 2.13 shows one of these passing Burketown in north Australia. It can be seen that (as in laboratory internal bores of strength 4 or greater) the Glory now resembles a turbulent atmospheric gravity current: however, the pressure pattern at the ground is still undular (Clarke *et al.*, 1981).

The origins of these disturbances are attributed to sea-breeze fronts interacting with an existing nocturnal inversion. The Morning Glory which arrives from the NE is essentially formed by the collision of the east- and west-coast sea breezes (Clarke, 1984; Noonan & Smith, 1986). Observations which can be interpreted as early stages in the formation of internal bores by sea-breeze fronts will be dealt with in Chapter 3.

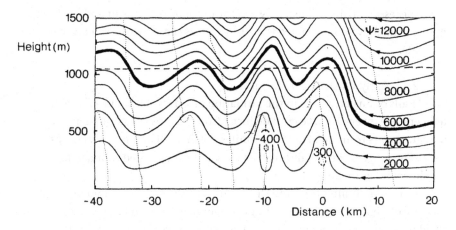

Figure 2.12 Relative streamlines, normal to the cloud lines, using seven pilot balloon ascents, of the Morning Glory, 4 October 1979. The thick line is the level of the top of the stable layer. Ψ is the value of the relative stream function in m s^{-1}.

Figure 2.13 The Morning Glory, turbulent form, passing Burketown, north Australia, 4 October 1979. (Courtesy of Derek Reid.)

2.4 Generation of bores

The most probable generation mechanism for an atmospheric bore is the disturbance of an existing stable layer by some kind of gravity current. The wide range throughout which a gravity current can generate a bore advancing in a two-layer system will be examined in detail in the laboratory experiments of Chapter 13. Provided that the depth of the gravity current is greater than that of the stable layer, it can be shown that the bores formed will be similar for a step density change, as in the experiments, and for a linear density gradient, which is more probable in the atmosphere.

Many of the gust fronts analysed by tower measurements have not fitted into the simple gravity current picture. In one series of 20 measurements (Goff, 1976a) the fronts often consisted of multiple surges and it was necessary to distinguish between the onset of the cold air and the high momentum flow which defined the gust front, as these two did not always coincide.

2.4.1 Early stages of bore formation

The details of an early stage in the formation of a disturbance in a low-level temperature inversion have been deduced from the pressure and temperature measurements from a 450 m tower 6 miles north of Oklahoma City (Marks, 1974). The observations made after the passage of an isolated thunderstorm are shown in figure 2.14.

Immediately after the wind shift at 1625 CST there was a rapid pressure increase of about 1 mb, but no substantial change in ground temperature. This rapid pressure rise is seen to be associated with a wave propagating on the

Figure 2.14 Observations made from an instrumented tower, showing an early stage in the formation of an atmospheric bore. (a) Ground level pressure. (b) Potential temperature from the tower data. (After Marks, 1974.)

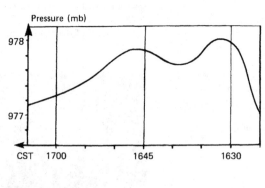

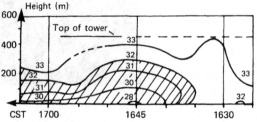

previously existing 100 m inversion. The temperature plot also shows a gravity current of cold air following the wave. Ground level pressure decreased as the wave passed over the tower but reached another peak at 1650 CST when the head of the gravity current was directly over the tower.

It appears from the temperature plot that a bore was in the first stages of generation when it passed the tower. It is estimated that the bore strength was about 3. The model described in Chapter 13 predicts a bore speed of $2(g'h)^{\frac{1}{2}}$ for this case, i.e. 8.1 m s^{-1}, which is close to the measured speed of the leading crest. The wavelength is deduced to have been about 7.3 km.

2.5 Development of a bore from a gravity current

Attempts have been made to explore the sequence of events in the formation of bores from gravity currents in the atmosphere. Because there is a wide band of flows that can occur between a pure bore and a pure gravity current, it is not always easy to distinguish a bore from a gravity current. Both may contain cool air, advected by a gravity current and cooled in ascent by a bore, but pronounced surface cooling is not a characteristic of bores; in fact, in a bore there may be a temperature rise due to downward mixing of warm air from the inversion layer.

The development of an undular bore in Oklahoma, USA, was followed by Fulton *et al.*, (1990) as it moved away from a cold thunderstorm outflow. Using doppler radar, a 440 m instrumented tower and a surface network the disturbance was measured as it passed the tower; the leading part of the current was found to have three periodic wind surges embedded in it, with a spacing of about 12 km.

Later measurements showed that the cold air was leaking out of the gravity current which decelerated, allowing the undular bore to propagate ahead into the prefrontal boundary layer. Figure 2.15 shows a schematic diagram of the disturbance based on the tower and doppler radar observations (figure 2.15a), and the speeds of the bore and the gravity current front based on data from the ground stations (figure 2.15b).

A feature of a study by Koch *et al.* (1991) of bore formation from Greenbelt, MA, USA, was the use of Raman lidar as well as radiosonde and surface data. A strong undular bore, which tripled the height of the surface inversion, was generated by a dissipating gravity current. The Raman lidar made it possible to study the intermediate structure between the gravity current and the undular bore, and the system showed a strong coherence over a distance of 180 km. Figure 2.16 shows a time–height cross section of potential temperature at the lidar site.

Not all gravity currents moving through a stable inversion layer will produce an undular bore. This was illustrated by Ralph *et al.* (1993) from field measurements in SW France in a study of a gravity current moving along a nocturnal inversion under an elevated neutral area. This work showed the usefulness of remote sensing such as radar and sodar in revealing the detailed structure of a gravity current and the scale of waves formed by such currents.

Waves in the inversion layer were calculated to have a speed of advance only half that of the gravity current front, so the case is supercritical; that is, a state in which any undulations which may have formed could not propagate ahead of the front.

The range of conditions in which forward propagating bores may be formed has been investigated in the laboratory and is described in Chapter 13.

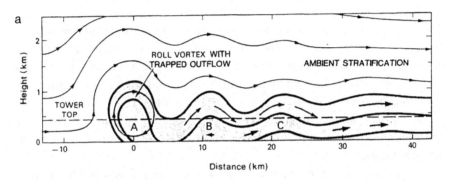

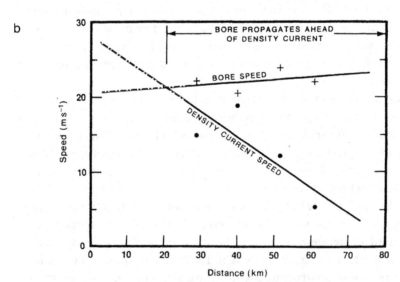

Figure 2.15 (a) Schematic diagram of cold outflow disturbance at Oklahoma, USA, based on tower and doppler radar observations. Three wavelike disturbances are marked at A, B and C. (b) Speeds of the bore front (+) and the gravity current front (●), based on four ground stations on a line normal to the front. (After Fulton *et al*, 1990, courtesy American Meteorological Society.)

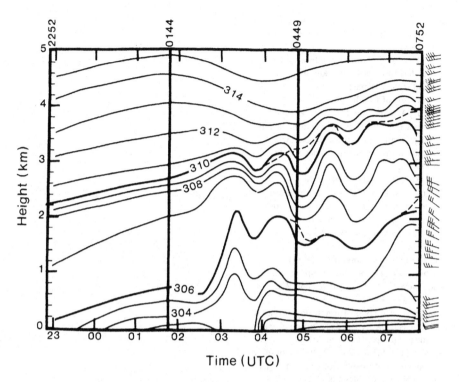

Figure 2.16 Cross-section of potential temperature of a structure intermediate between a gravity current and an undular bore. Measurements were made by lidar, radiosonde and ground data. (After Koch *et al.*, 1991; courtesy American Meteorological Society.)

Bibliography

Carbone, R.E. (1982). A severe frontal rain band. Part 1. *J. Atmos. Sci.*, **391**, 258–79.

Christie, D.R., Muirhead, K.J. & Hales, A.L. (1979). Intrusive density flows in the lower troposphere: a source of atmospheric solitons. *J. Atmos. Sci.*, **84**, 4050–70.

Clarke, R.H. (1984). Colliding sea-breezes and the creation of internal atmospheric bore waves: two-dimensional numerical studies. *Austr. Meteorol. Mag.*, **31**, 207–26.

Clarke, R.H., Smith R.K. & Reid, D.G. (1981). The Morning Glory of the Gulf of Carpentaria: an atmospheric undular bore. *Mon. Wea. Rev.*, **109**, 1726–50.

Fulton, R., Zrnic, D.S. & Doviak, R.J. (1990). Initiation of a solitary wave family in the demise of a nocturnal thunderstorm density current. *J. Atmos. Sci.*, **47**, 319–37.

Garratt, J.R. (1984). Cold front and dust storms during the Australian summer, 1982–3. *Weather*, **39**, 98–103.

Goff, R.C. (1976a). *Thunderstorm-Outflow Kinematics and Dynamics.* NOAA Technical Memo. ERL NSSL-75. 69 pp.

Goff, R.C. (1976b). Vertical structure of thunderstorm outflows. *Mon. Wea. Rev.*, **104**, 1429–40.

Hall, F.F., Neff, W.D. & Frazier, T.V. (1976). Wind shear observations in thunderstorm density currents. *Nature*, **264**, 408–11.

Idso, S.B. *et al.* (1982). American haboob. *Bull. Am. Met. Soc.*, **53**, 930–35.

Kharitanova, S.S. (1969). Dust content of the air during dust storms in the Mangyshlak Peninsula. *Met. i. Giaro.*, **5**, 87–8.

Koch, S.E., Dorian, P.B., Ferrare, R., Melfi, S.H., Skillman, W.C. & Whiteman, D. (1991). Structure of an internal bore and dissipating gravity current as revealed by Raman lidar. *Mon. Wea. Rev.*, **119**, 857–87.

Lawson, T.J. (1971). Haboob structure at Khartoum. *Weather*, **26**, 110–12.

McAllister, L.G. (1968). Acoustic sounding of the lower troposphere. *J. Atmos. Terr. Phys.*, **30**, 1439–40.

Marks, J.R. (1974). *Acoustic Radar Investigations of Boundary Layer Phenomena*. Report NAS8–28659. Department of Meteorology, University of Oklahoma.

Noonan, J.A. & Smith, R.K. (1986). Sea breeze circulations over Cape York Peninsula and the generation of Gulf of Carpentaria cloudlike disturbances. *J. Atmos. Sci.*, **43**, 1679–95.

Ralph, F.M., Mazaudier, Crochet, M. & Venkateswaran, S.V. (1993). Doppler sodar and radar wind-profiler observations of gravity-wave activity associated with a gravity current. *Mon. Wea. Rev.*, **121**, 443–63.

Schreffler, J.H. & Binkowski, F.S. (1981). Observations of pressure jump lines in the Midwest, 10–12 August 1976. *Mon. Wea. Rev.*, **109**, 1713–25.

Shimizu, H., *et al.* (1985). Large scale laser radar for measuring aerosol distribution over a wide range. *Appl. Optics*, **24**, 617–26.

Simpson, J.E. & Britter, R.E. (1980). A laboratory model of an atmospheric mesofront. *Q. J. R. Meteorol. Soc.*, **106**, 485–500.

Smith, R.K., Crook, N. & Roff, G. (1982). The morning glory: an extraordinary atmospheric undular bore. *Q. J. R. Meteorol. Soc.*, **108**, 937–56.

Sutton, L.J. (1951). Haboobs. *Q. J. R. Meteorol. Soc.*, **57**, 143–61.

Tepper, M. (1950). A proposed mechanism of squall lines: the pressure jump line. *J. Meteorol.*, **7**, 21–9.

Wakimoto, R.M. (1982). The life cycle of thunderstorm gust fronts. *Mon. Wea. Rev.*, **110**, 1060–82.

3   Sea-breeze fronts

During a day of fine weather the sea breeze blows from the sea towards the land:
it is caused by diurnal temperature differences between land and sea. When the
sun shines, the sea surface temperature changes very little, but the land becomes
hotter and convection currents distribute heat through a layer of air one or two
kilometres above the ground. The sideways expansion of a column of air above
the land (see figure 3.1) produces changes in pressure which are transmitted
sideways at the speed of sound. The resulting pressure difference at low levels is
responsible for the onset of the sea breeze.

The sea breeze caused by this pressure gradient at low levels is usually only
three or four hundred metres deep but the distance that the sea breeze blows
inland may extend many tens of kilometres during the day. On a calm day the
inland boundary of this spreading cool air is initially quite diffuse, extending
over several kilometres.

On days when the sea breeze meets a steady opposing wind blowing
toward the sea, a sharp boundary forms between the land-air and the cooler sea-
air. A sea-breeze front develops, which has all the characteristics of a gravity
current of cold, dense air. It is marked by a gust front of cold air and is very like

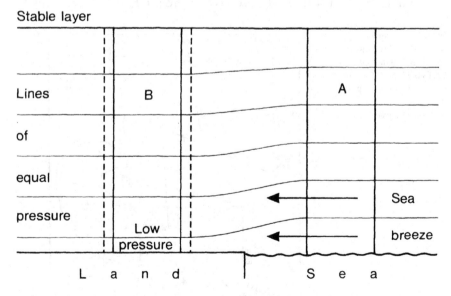

Figure 3.1 The development of the pressure field which gives rise
to the sea breeze at lower levels. Columns of air, A, above the sea,
B, above the land.

the gravity current formed by a cold outflow from a thunderstorm, as described in Chapter 2.

Several of the earliest aerial investigations of these sea-breeze fronts were made by meteorologists flying gliders, using the strong rising air at the front (Wallington, 1959). The sea breeze is significant to glider pilots in flight-planning because the sea-breeze front provides upcurrents and marks the seaward limit of good soaring conditions.

The 'sea-air' may extend over a large part of England on a fine summer day; an example of this is given in figure 3.2 of a day of light winds, mainly from the NW. This shows the sea-breeze situation on a good day for gliding and one of well-developed sea breezes. Sea-breeze fronts had formed parallel to several coasts and by the time shown had travelled nearly 50 km. The map also depicts a number of memorable soaring flights made above the strongly convective countryside where the stable sea breeze inversion layer had not yet encroached.

3.1 Formation and structure of sea-breeze fronts

Any pre-existing wind system has a great influence on the development of sea-breeze fronts. When there is already a wind blowing from the land towards the sea, opposing the sea breeze, these converging winds may develop a sharp front a few kilometres out to sea; whether it will begin to travel inland later during the day depends on the balance between the synoptic wind and the temperature difference between land and sea. The theoretical dimensionless number determining the onset of the sea breeze has been given as $F_s = U^2/(C_p \Delta T)$ (Biggs &

Figure 3.2 The spread of the sea breeze in England on a day of light winds (at 1800 GMT on 9 June 1968). The lines show tracks of glider flights in the area which remained convective.

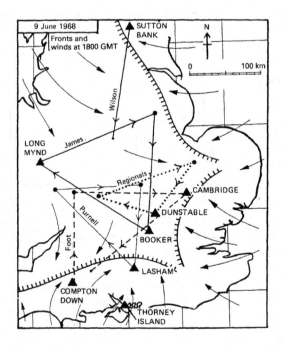

Graves, 1962), where U is the forecast offshore wind, C_p is the specific heat of air at constant pressure and ΔT the forecast temperature difference between land and sea. This is borne out by observations at Thorney Island in southern England (Watts, 1955), and at Lake Erie and Lake Michigan when forecasting the lake breeze. $F_s = 10$ is found to be the critical value above which no sea breeze reaches the shore.

On mornings of calm or very light wind, sharp sea-breeze fronts are not usually detectable near the coast, but may develop and accelerate inland after the time of maximum heating. Although on these days sea-breeze air may spread inland early in the morning, as it moves over the heated land its properties may be considerably modified, and by midday the temperature of the foremost sea-air is not very different from that of the land-air (Simpson, 1994). At this time any attempts to detect a front from aerial measurements have shown the changes of temperature and humidity to be spread over a distance of several kilometres.

An example of this situation is illustrated in figure 3.3, which shows the potential temperature gradient in the sea-air in the early afternoon, at 1500 GMT. The horizontal gradient of temperature is seen to be about $\frac{1}{4}\,°$C per km, extending over at least 4 km (Reible *et al.*, 1993).

Two hours later, as shown in figure 3.3b, the vertical mixing had decreased enough for the horizontal temperature gradient to steepen and to develop a sharp sea-breeze front.

In general, the formation of a sea-breeze front depends on the balance between converging winds, which act to form a front, and the vertical mixing over the land, which acts to prevent its formation. It is even possible for a sea-breeze front to form on a day of light onshore wind, provided that the land–sea temperature difference is large enough. Since the wind convergence is small a front will not form until late in the day, 50 km or more from the coast.

3.1.1 Inland advance of sea-breeze fronts

The daily progress of the sea breeze inland in southern England has been monitored in a project based at the Lasham Gliding Centre. During a period of ten years, the sea breeze passed Lasham, which is 45 km from the coast, on as many as 76 days (Simpson *et al.*, 1977).

Figure 3.4 shows the progress of the sea breeze inland on 14 June 1973, a summer day of light winds, typical of many observations of deep inland penetration. Data from 21 ground stations were backed up by seven balloon ascents and two aircraft flights. The times of passage shown here differ by less than an hour from the mean times of the 76 cases, although not all were traced so far inland after passing Lasham. In the earlier stage, with no offshore wind to tighten up a front, the leading edge of the sea breeze was diffuse and extended over a kilometre or more. Later during the day, by 1600 GMT, a distinct front had developed, the position of which could be identified by very distinctive cloud forms, with base level much lower than that of the cumulus cloud further inland.

Observations of the structure of many sea-breeze fronts have confirmed that they behave like gravity currents of dense air. Typical measurements made from an instrumented light aircraft near such a front are shown in figure 3.5. The aircraft used was a Falke 'motor-glider', which can be flown slowly and which has many features similar to a glider. It carried a data-logging system which recorded height, airspeed, rate of climb, air temperature and water vapour density every 1.6 seconds, or every 50 m along the flight path. This plot of the humidity cross-section is built up from six horizontal traverses, made between 1820 and 1900 GMT, and shows the pattern near a sea-breeze front along a distance of 2 km, up to a height of 1 km.

The sea breeze was 400 m deep, but the observations show moist air

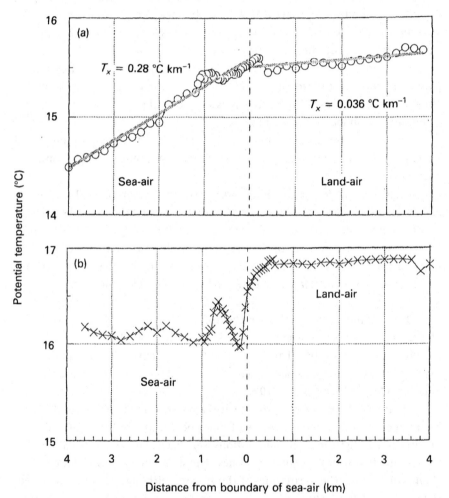

Figure 3.3 Sea-breeze frontogenesis on a calm day (14 June 1973). Horizontal temperature profiles are measured at a height of 300 m through the boundary of the sea air. T_x is the temperature gradient in degrees per kilometre. (a) 1500–1605 GMT, showing a change to an almost horizontal temperature gradient. (b) 1730–1805 GMT, showing a sharp sea-breeze front.

reaching about twice this height, and the form suggests the formation of billows as seen in laboratory flows. Further similarities were shown by measurements of four other fronts under similar conditions, which showed the initial height of moist sea-air to be twice as great as that of the sea breeze 10 km behind the front (Simpson *et al.*, 1977).

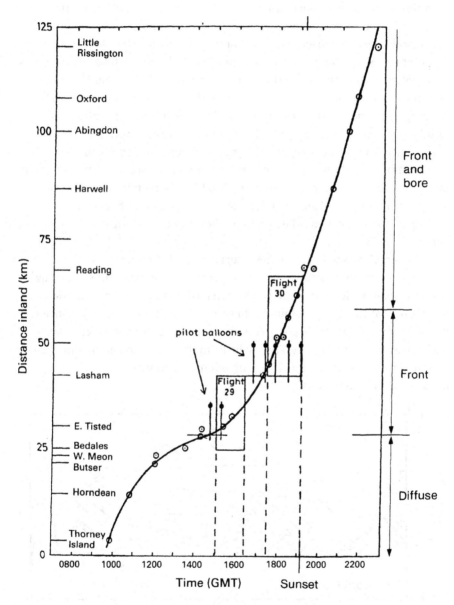

Figure 3.4 The progress of the sea breeze inland on 14 June 1973, a sunny day of light winds. Before 1600 GMT the leading edge was diffuse; later, a distinct front had developed. The data points show the time of arrival of the sea breeze as detected at ground stations measuring temperature, relative humidity and wind direction. Also marked are times and positions of pilot balloon ascents and two flights made in an instrumented light aeroplane.

3.1.2 Cloud forms at a sea-breeze front

The presence of a sea-breeze front can often be deduced from the ground by the formation of distinctive forms of cloud. On a clear day the rising air at the sea-breeze front may reach condensation height and form a line of cloud in an otherwise clear sky. On a day in which the sky is full of small fair-weather cumulus, the front may make its presence visible both as a seaward limit to cumulus formation and by fragments of cloud with a much lower cloud-base. These clouds form as ragged veils or curtains and can be seen to be rising rapidly (in other respects they have the appearance of dissolving cumulus cloud). A view of a sea-breeze front seen from the air, flying just beneath the cloud-base of the cumulus clouds, is shown in figure 3.6. Three short lines of cloud can be seen forming along the upper surface of the sea-breeze front, which is also made visible by the haze it has brought inland from the coast at Southampton and Portsmouth. The profile is typical of a gravity current moving in calm surroundings and three advancing lobes can be seen, with a separation of about one kilometre. Viewed from the side the clouds can be seen to be sloping, but viewed from the ground ahead of the front the slope cannot be detected and the ragged cloud can merely be seen hanging below the base of the cumulus clouds.

Glider pilots investigating these 'curtain clouds' found that they marked a very narrow band of the strongest uplift. This rising air was sometimes only a few wing-spans wide and the mystery was that the penalty of flying on the 'wrong side' of this curtain cloud could be a very sudden area of turbulence and strong downcurrents. The glider pilots' findings are now seen to be consistent with the gravity current story. Those who fly in the upper part of an atmospheric gravity current can find strong rising air, but must beware of fearsome downcurrents in the billows.

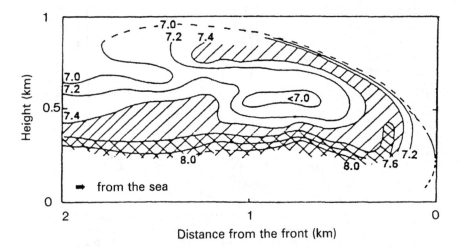

Figure 3.5 Humidity cross-section of a sea-breeze front, showing isopleths of mixing ratio in g kg^{-1}. Diagonal shading=7.4–7.6 g kg^{-1}, cross hatching=>7.6 g kg^{-1}.

3.2 Internal bores formed by the sea breeze

One part of the world in which the progress of the sea breeze has been extensively studied is South Australia, where sea-breeze effects have been followed more than 300 km inland (Clarke, 1965). In the course of these studies it was found that the leading edge of the sea breeze at sunset often appeared to be in the form of a 'vortex' which sometimes separated from the following flow. Glider pilots flying at the sea breeze in southern England also noticed a marked change in the structure of the sea breeze upcurrent near sunset. Instead of the rough, uncertain strip of lift found near curtain clouds earlier in the day, the rising air had spread to cover a much larger area and had also become very smooth. The characteristics had changed to those found by glider pilots when soaring in rising air displaced by atmospheric waves.

'Sea-breeze vortices' were subsequently found in England, and sometimes as many as three successive ones were traced, moving inland at 4 m s^{-1} with a spacing of about 10 km. The vortices can now be interpreted as being the early stages in the generation of an atmospheric bore by the movement of the sea-breeze front in the stable layer formed by the nocturnal inversion.

An early stage of the formation of an internal bore in the inversion has been measured north of Lasham and appears in figure 3.7. This shows the streamlines at the sea-breeze front just before sunset on 14 June 1973. It is clear that although the sea-breeze front in the early afternoon had been of gravity current form, the foremost part was by then almost separated from the following

Figure 3.6 Clouds and haze at a sea-breeze front near Selborne, UK, 9 June 1968. (Courtesy of H. Howitt.)

flow. The form resembles the first wave of an undular bore beginning to develop around the head of a gravity current as it moves through a stable layer, as illustrated in Chapter 13 in figure 13.7. The length of this section gives an estimate of 7 km for the wavelength of the undular bore.

This 'sea-breeze bore' on 14 June 1973 was recorded at six more stations as it progressed inland. In the wind record at Harwell, 90 km from the coast, shown in figure 3.8, signs of waves are visible up to three hours after the first wave disturbance. The time for a wave to pass was about three-quarters of an hour; at the measured velocity of 12 km h^{-1} this gives a wavelength of 9 km.

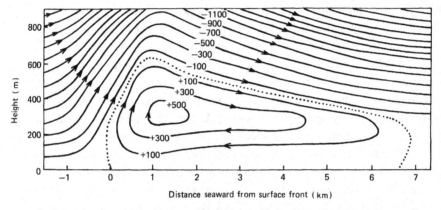

Figure 3.7 Streamlines at the sea-breeze front on 14 June 1973, just before sunset, showing the incipient bore.

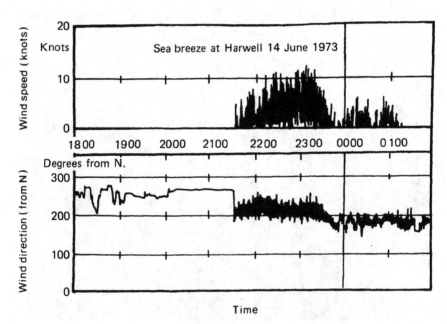

Figure 3.8 Wind strength and direction at Harwell on 14 June 1973. The oscillations are associated with the bore generated by the sea breeze.

3.3 Sea breeze and pollution

One of the most important practical applications of the study of sea-breeze fronts is the effect on the distribution of airborne pollution. At first thought it might seem that the effect of the sea breeze is to enhance the dispersion of pollution, but actually it is found that it may concentrate undesirable aerosols. Field work using tetroons, which are constant-pressure balloons used as tracers, has confirmed the existence of recirculation patterns in the sea breeze. Such recycling can concentrate airborne pollutants and make possible undesirable chemical changes. During extensive measurements of the sea-breeze flow in the neighbourhood of Chicago (Lyons & Olsson, 1972), one of the tetroons was found to make a complete circulation in 3.5 hours.

At the coastal city of Los Angeles, under typical sea-breeze conditions a shallow layer of cool moist marine air flows inland over the urban area and becomes heavily contaminated with the ingredients of photochemical smog. About a dozen times a year the sea breeze reaches Riverside, about 60 miles from the origin of the smog, as a sharp sea-breeze front between the polluted marine air and the clean desert air. One such front was photographed as it passed across Riverside in the early afternoon of 16 March 1972, and is shown in figure 3.9 (Stephens, 1975). The photograph was taken with a polaroid filter to darken the sky and increase its contrast with the cloud of polluted air. The

Figure 3.9 The front of a sea breeze polluted by photochemical smog, passing Riverside, California, 16 March 1972. (Courtesy of G.R. Stephens.)

visible part of the smog cloud was about 1000 m deep and consisted of the aerosol; it could not have been fog, since the relative humidity was only about 30%.

The characteristic feature of photochemical smog is its oxidising power, and part of the oxidant record for this day is shown in figure 3.10. When the smog front passed the air-monitoring station the rise in oxidising power was so rapid that its measurement was limited by the response time of the instrument. It can be seen that the value rose from 0.06 parts per million to a peak value of 0.43 ppm. Incidentally, the US federal air quality standard is 0.08 ppm.

A similar smog front used to be common in the Middlesborough district on the NE coast of England. During the morning a sharp smog front, very clearly marked by smoke haze, often crossed the town, reducing visibility to 100 m or less. Some ragged cumulus sometimes appeared at the front, with a condensation level much lower than that of the main cumulus clouds. However,

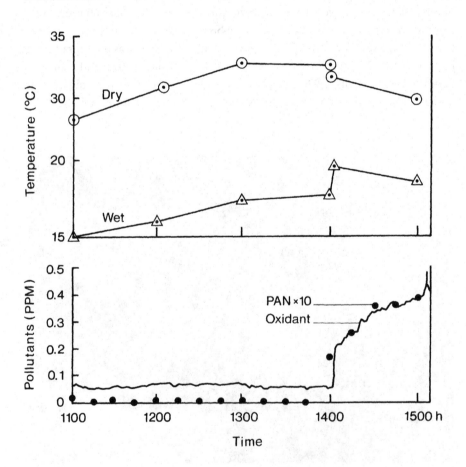

Figure 3.10 Records of wet and dry temperature and chemical concentrations at the front passing Riverside, California, 16 March 1972 (PAN is peroxyacetyl nitrate). (After Stephens, 1975.)

with the reduction of pollutants in the air, this phenomenon is now much less common.

It appears that the air freshening that has been believed to accompany sea breezes may be in some cases purely illusory, and pollutants may be concentrated continually by the sea breeze, so that exceptionally high values can occur inland at sea-breeze fronts.

Some values of the frequency of inland penetration of sea breezes have been measured in southern England. The chart in figure 3.11 shows a summary made from continuous monitoring from 1962–1973, where the logarithmic plot shows that the frequency falls off quickly with distance from the coast, reducing to less than one a year beyond 100 km inland.

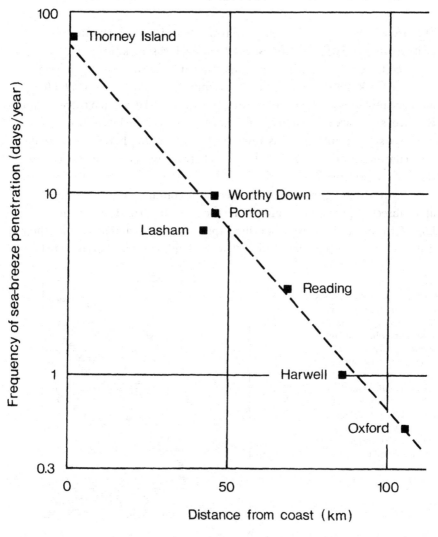

Figure 3.11 The frequency of penetration of the sea breeze to points inland from the south coast of England, averaged over a period of 12 years.

3.3.1 Sea breeze remote sensing by lidar

The name 'lidar' has been given to the technique of remote sensing in the atmosphere by the use of light, usually a laser beam. Using the presence of aerosols in the atmosphere as tracers, lidar has been used to measure the structure of sea-breeze fronts.

The lidar display shown in figure 3.12 is the vertical cross section of a sea-breeze front, showing the concentration distribution of aerosols (Nakane & Sasano, 1986). The observations show that the inflowing sea breeze was about 300 m deep, with a much deeper zone of turbulent mixing above it.

3.4 Birds and insects at sea-breeze fronts

The bird which has become especially associated with sea-breeze fronts is the swift (*Apus apus*). Swifts are almost wholly aerial birds, picking up their livelihood, even their nesting materials, from airborne flotsam. It has been recorded (Lack, 1956) that on a fine day a single swift can catch up to 1000 airborne insects every hour. To continue at this rate all day when there are young to be fed, swifts are obliged to fly in places where there are lots of insects. Glider pilots see swifts hunting for insects in thermals, and they have been noted in high concentrations flying along the line of strongest upcurrent at sea-breeze fronts (Simpson, 1967).

Swifts catch a great variety of insects, the commonest being aphids and allied insects. Aphids, which are among the most important agricultural pests in England, are only weak flyers but they exploit large-scale air currents for their distribution. As a sea-breeze front advances inland, swarms of aphids are lifted

Figure 3.12 Analysis of a vertical cross-section of a sea-breeze front seen by lidar. This observation was made at 1740 hours, when the front, moving at 3.3 m s⁻¹, had reached a point 45 km inland from Tokyo, Japan. The flow can be classified into three regimes: 1, the inflowing sea breeze; 2, the returning mixed flow; and 3, the ambient wind. (After Nakane & Sasano, 1986.)

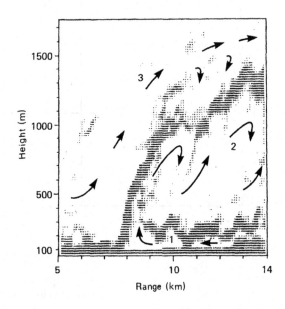

into the band of rising air ahead of the front, and swifts soaring at the front can find a continuous supply of their food.

3.4.1 Radar observations of sea-breeze fronts

The progress of an intense sea-breeze front moving inland in SE England was first followed inland in 1961 (Eastwood & Rider, 1961). On this occasion it was concluded that both moisture gradients and airborne swifts feeding at the front contributed to the radar returns. Later work established the presence of 50 times the minimum number of swifts calculated to give the observed echo-strength. Figure 3.13 shows a radar display of a sea-breeze front made during these observations, on 2 June 1966.

3.4.2 Insects at sea-breeze fronts

The radar echoes shown in figure 3.13 were made using a very high power radar, working at 23 cm wavelength: however, the development of radar using much shorter wavelengths has made it possible to detect insects. They have been detected in clouds, and it has been possible to detect and count individual insects from airborne radar.

Figure 3.13 Sea-breeze fronts in SE England outlined by radar, at 1530 GMT, 2 June 1966. The large dots are echoes from aeroplanes, and the smaller ones are from groups of birds. The Marconi radar station at Chelmsford in SE England is in the centre of the circle, which is of 100 km radius, and the south coast is marked as far west as the Isle of Wight. The arrow points to the line of a sea-breeze front which had moved inland from the south coast, and along which swifts were being counted. Another sea-breeze front can also be seen about 30 km inland from the east coast at the top of the picture. (Courtesy of Marconi Limited.)

Figure 3.14 shows the aerial density of flying moths seen by 3 cm radar from an aeroplane flying across a sea-breeze front (Schaefer, 1979). This airborne radar traverse made in New Brunswick, Canada, shows the relative concentrations of spruce budworm moths (*Choristoneura fumiferana*) up to a height of 500 m. The distance extends from 3 km ahead of the front to as far as 14 km behind the front. It can be seen that the shape of the nose and form of the top of the head resemble the form of a gravity current.

3.4.3 Insects at other small-scale fronts

Studies of the distribution of insect pests in Africa, particularly of the desert locust (*Schistocerca gregaria*), have made clear the importance of weather systems in their life histories. For example, as a swarm of locusts approaches a coastline it has a good chance of meeting the sea breeze, which will prevent any possibility of being blown out to sea. The effect of sea-breeze fronts in concentrating the swarms also has an important influence on the behaviour of the insects.

Another important flying insect pest is the African armyworm moth (*Spodoptera exempta*), which flies at night. Radar detection of these insects has been used to trace the airflow in the atmosphere, particularly at fronts. Pedgley *et al.* (1982) used 3 cm radar during field work in Kenya to monitor individual moths flying within a distance of about 2 km. A cloud of moths, not so dense as to cause overlapping echoes, can reveal wind systems on a scale of a kilometre or so by means of time-lapse photography of the radar screen, using one frame for each 3-second radar sweep. As the wind-shift line passed, marking the front of the cold outflow, it was possible to build up a vertical cross-section of echo motions. The outflow was clearly overtaking the wind shift, as in the typical head of a gravity current.

There was no evidence that the strong concentration of moths was due to

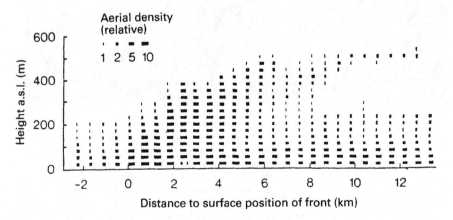

Figure 3.14 Density of flying moths at a sea-breeze front, seen from an airborne radar traverse, New Brunswick, Canada, 10 July 1976, at 2337 h. (Courtesy of K.A. Allsopp.)

Figure 3.15 The form of a cold outflow outlined by flying locusts, taken at Hargeisa in the Somali Republic, 3 August, 1960. (Courtesy of A.J. Wood.)

lifting of moths already flying densely near the ground, and it seemed probable that the moths were being concentrated by the convergent winds. The dramatic density increase that was observed needed only a descent of insects behind the front.

Locusts fly during the day, so the patterns formed by their swarms can be seen and photographed. Figure 3.15 gives a striking view of a cold outflow outlined by flying locusts. The pattern marked is similar to that shown in the radar measurements. The locusts can be seen moving in from the right and being swept upwards and backwards as they reached the front.

Bibliography

Biggs, W.G. & Graves, M.E. (1962). A lake breeze index. *J. Appl. Meteorol.*, **1**, 474–80.

Clarke, R.H. (1955). Some observations and comments on the sea breeze. *Austr. Meteorol. Mag.*, **11**, 47–68.

Clarke, R.H. (1965). Horizontal mesoscale vortices in the atmosphere. *Austr. Meteorol. Mag.*, **50**, 1–25.

Eastwood, E. & Rider, G.C. (1961). A radar observation of a sea-breeze front. *Nature*, **189**, 978–80.

Lack, D. (1956). *Swifts in a Tower*. London: Methuen. 239 pp.

Lyons, W.A. & Olsson, L.E. (1972). Mesoscale air pollution transport in the Chicago lake breeze. *J. Air Pollut. Control Assoc.*, **22**, 876–81.

Nakane, H. & Sasano, Y. (1986). Structure of a sea-breeze front revealed by a scanning Lidar observation. *J. Meteorol. Soc. Japan Ser. II*, **64**, (5), 787–92.

Pedgley, D.E., Reynolds, D.R., Riley, J.R. & Tucker, M.R. (1982). Flying insects reveal small-scale wind systems. *Weather*, **37**, 295–306.

Reible, D.D., Simpson, J.E. & Linden, P.F. (1993). Sea breeze and gravity current frontogenesis. *Q. J. R. Meteorol. Soc.*, **119**, 1–16.

Schaefer, G.W. (1979). An airborne radar technique for the investigation and control of migrating insect pests. *Phil. Trans. R. Soc. Lond.*, **B287**, 459–65.

Simpson, J.E. (1967). Aerial and radar observations of some sea-breeze fronts. *Weather*, **22**, 306–17.

Simpson, J.E. (1994). *Sea Breeze and Local Wind.*
    Cambridge University Press, 234 pp.
Simpson, J.E., Mansfield, D.S. & Milford,
    J.R. (1977). Inland penetration of sea-
    breeze fronts, *Q. J. R. Meteorol. Soc.*, **103**, 47–76.
Stephens, E.R. (1975). Chemistry and
    meteorology in an air pollution episode. *J. Air
    Pollut. Control Assoc.*, **25**, 521–4.
Wallington, C.E. (1959). The structure of the sea-
    breeze front as revealed by gliding flights.
    Weather, **14**, 263–70.
Watts, A. (1955). Sea-breeze at Thorney Island,
    Meteorol. Mag., **84**, 42–8.

4   Gravity currents in satellite imagery

4.1 Use of satellites

Many manifestations of gravity currents in the atmosphere and on the surface of the ocean may be observed in images captured by either geo-stationary or polar orbiting satellites.

4.1.1 Geo-synchronous satellites

Geo-synchronous, or geostationary, satellites are placed above the equator, at a height of 35 800 km, where their orbiting speed keeps them above the same point and they are able to monitor continuously almost one half of the earth's surface. One great advantage of a geostationary satellite is that it produces a regular series of images which then can be recorded to give a motion sequence showing cloud development.

4.1.2 Polar orbiting satellites

These orbit about 840 km above the earth's surface with an orbital period of approximately 100 minutes. As the earth rotates beneath, the whole of the earth's surface becomes visible in 24 hours, provided the width of view is sufficiently large.

4.1.3 Observational channels

Two main channels are employed in satellite imagery, the visible and the infra-red. The channel in the visible spectrum works like ordinary photography using radiation reflected from the earth. The infra-red detector measures 'black-body' radiation, which is emitted by all objects day and night and gives a measure of their absolute temperature. Since black-body radiation is continuously emitted by all objects, the infra-red channel gives a direct measurement of the temperature of all points in view, through both day and night.

Since the temperature profile in the atmosphere is known, one can deduce the altitude of various cloud structures from infra-red images. By convention, the lightest parts of the image represent low temperatures and darker parts the high temperatures.

Figure 4.1 shows an infra-red image from the geostationary Meteosat 5 over the Greenwich meridian on a September morning. The brightest points are the tops of storm clouds, several of which can be seen over Africa, with temperatures of about −40 °C. The darkest points are on the ground in Arabia at about +40 °C.

The storm off West Africa, which was being monitored as an incipient hurricane, is surrounded by a ring of cloud of about 300 km radius, and has two spiral fingers of cloud, suggesting that strong rotation is already present.

4.2 Rope clouds

Images from weather satellites on both visible and infra-red channels often show thin 'rope clouds', or 'arc clouds'. Motion sequences from geostationary satellites have followed these clouds for several hours while they travel 100 kilometres or more.

4.2.1 Thunderstorm outflows

Motion sequences have made it possible to follow rope clouds as they spread radially from developing thunderstorms; they may maintain their identity for several hours after the convection has dispersed, and travel distances of 200 km or more (Purdom, 1973). The rope cloud around the storm complex in figure 4.1 had spread 400 km from its centre.

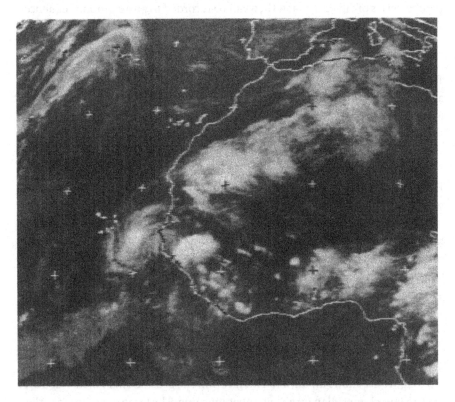

Figure 4.1 Infra-red image from the geostationary satellite Meteosat 5, above the Greenwich meridian. The large storm near the West African coast has a ring of rope cloud around it, and some spiral arms.

Surface measurements backed up by radar and ground observations have
confirmed the similarity of rope clouds to those marking squall lines at
thunderstorm outflows (Matthews, 1981). Figure 4.2a is one frame from a
sequence showing a rope cloud which has moved north and east from the storm.
Figure 4.2b, a radar plan view, shows the same arc cloud structure propagating
away from the storm centre, G. Figure 4.2c, composed from aircraft observations
on a flight through the structure illustrated in figure 4.2a, provides an
illustration of the airflow and cloud formations.

'Mergers' of two or more rope clouds from different sources have been
observed. These indicate where new showers are most likely to form, and so may
be used in forecasting (Purdom, 1976).

4.2.2 Morning Glory

The 'Morning Glory' is a distinctive cloud form observed in the Gulf of
Carpentaria in Australia. This undular bore has been described in section 2.3.1
with photographs taken both from the ground and from aircraft. The line of
cloud marking the Morning Glory has been detected in satellite imagery and
figure 4.3 (Smith *et al.*, 1995) shows the bore extending over a length of 300 km as
two roll clouds separated by about 20 km.

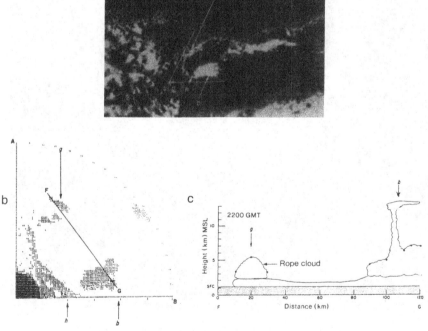

Figure 4.2 Rope cloud moving north and east from a storm. (a)
Satellite imagery. (b) PPI Radar echoes. (c) Airflow and cloud
formations along the line FG. (Matthews, 1981, courtesy of the
American Meteorological Society.)

4.2.3 Low-level air boundaries in the Mediterranean

Rope clouds marking low-level boundaries between air masses of different origin have been described by Scorer (1986, 1990): an example of this in the Mediterranean is given in figure 4.4. This image was taken above Sardinia; the rope cloud extending roughly from Barcelona to Nice indicates that there is no mixing at the edge of the moister air to the north with the drier air to the south. This cloud acts as an indicator of a shallow cold front.

4.2.4 Tehuantepecer: cold flow from the Gulf of Mexico

Another localised rope cloud of regular occurrence is formed by the 'Tehuantepecer'. When the cold air mass in the Gulf of Mexico is deep enough it spills over the isthmus and rushes down the western slopes of the mountains into the Pacific. This causes a strong wind often encountered during the winter by ships travelling from the Panama Canal through the Gulf of Tehuantepec.

Satellite imagery in figure 4.5a (Parmenter, 1970) shows a circular rope line marking the leading edge of the cold air outbreak through the narrow pass in the mountains. The passage of cold air was well established by the time the ESSA 9 photograph was taken and the movement of this rope cloud was measured from later pictures, shown in figure 4.5b. The rate of advance of this squall-line has been shown to agree with gravity current theory.

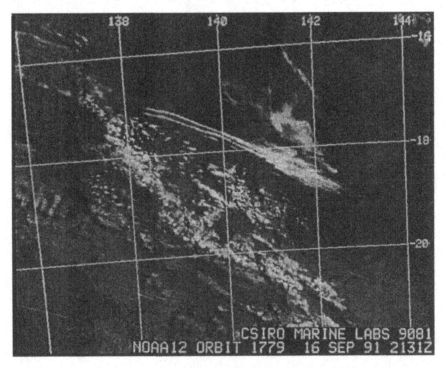

Figure 4.3 The 'Morning Glory' cloud in Australia. This is an undular bore, with two roll clouds 20 km apart. (Courtesy R. Smith.)

4.2.5 Rope clouds across the Pacific

The image in figure 4.6a marks the leading edge of a gravity current of cold air which travelled from Siberia across Japan to the Pacific. The air was 10 °C colder than the sea surface: during the consequent heating instability patterns can be seen for hundreds of kilometres behind the front. Figure 4.6b gives traces at 6 hour intervals, showing the progress towards the south of about 30 km in each hour, an average speed of 8.3 m s^{-1}.

4.3 Ropes and cold fronts

Long, thin rope clouds are frequently found immediately parallel to synoptic cold fronts, at typical heights of 2500–3000 m. Although such rope clouds have more often been seen over tropical oceans, they also occur over land in the SE United States.

4.3.1 The 'cold outflow' model

A rope cloud at a cold front across Florida is shown in figure 4.7. The rope cloud moved close to the cold front, which had heavy precipitation near its leading

Figure 4.4 Rope cloud marking low-level boundaries between air masses in the Mediterranean. The rope, which extends from Barcelona to Nice, shows that there is no mixing at the edge of the moister air to the north. (Courtesy of the University of Dundee.)

edge. Measurements of this rope cloud using a 150 m instrumented tower
(Seitter & Muench, 1985) showed it to be very similar to those observed at gust
fronts from thunderstorm outflows. However, not all rope clouds at cold fronts fit
into this model, because some appear independently of any precipitation lines.

4.3.2 The 'self-sharpening front' model

Figure 4.8 illustrates two possible formation processes for rope clouds at cold
fronts.

It seems certain that moist cold fronts with precipitation can produce
gravity current flows which form rope clouds close to the leading edge of the
front. Sometimes these outflows are capable of initiating bores in a stable layer
(Rottman & Simpson 1989), and so provide another potential source of rope
clouds.

Evidence has also been given by Shapiro *et al.* (1985) of the emergence of
a gravity current from a cold front in the absence of precipitation, by way of
'cross-frontal collapse'. This self-sharpening process is shown in figure 4.8b. As in

Figure 4.5
(a) The 'Tehuantepecer'
photographed on 3 February
1970. F points to the edge of
the frontal cloud band along
the Sierra Madre Range.
G shows the cloud which
marked the forward edge of
cold air. (b) 'Tehuantepecer',
showing cold air from the
Gulf of Mexico moving
through a narrow pass in the
mountains. Contours of the
mountains are shown shaded.
(From Parmenter, 1970,
courtesy of the American
Meteorological Society.)

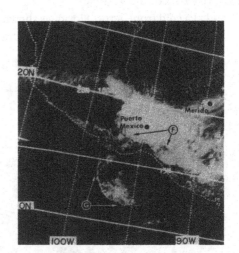

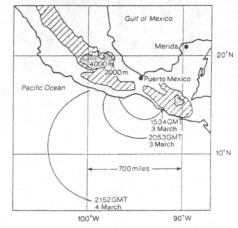

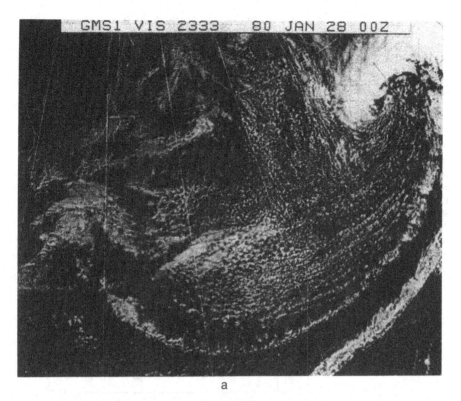

a

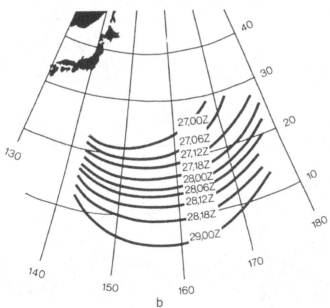

b

Figure 4.6 Rope cloud at the boundary of cold air moving from the SE of Japan, covering 30 degrees of longitude. (a) Image from the Japanese geostationary satellite GMS1, on 28 January 1980. (b) Traces at 6 hour intervals, showing the rope cloud moving at 8.3 m s^{-1}. (Courtesy of R. Kimura.)

the case of the leading edge of a thunderstorm outflow, the leading edge of this gravity current may be marked by a rope cloud. Whether or not cold fronts often sharpen up to develop the local character of a gravity current is still an open question (Smith & Reeder, 1988). More data may enable this question to be answered.

Figure 4.7 Rope cloud at a cold front over Florida. (a) Visible satellite image at 1300 GMT. (b) Position of the front at 1200 GMT. (After Seitter & Muench, 1985, courtesy of the American Meteorological Society.)

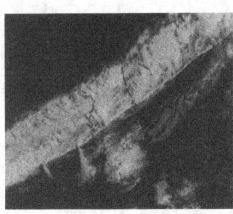

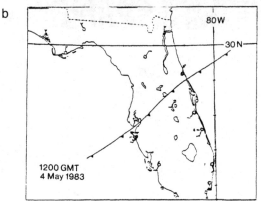

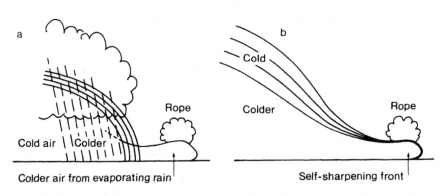

Figure 4.8 Possible formation processes for rope clouds at a cold front. (a) Colder air from evaporating rain, (b) 'self-sharpening' front.

4.4 Ropes in the sky

The name 'Ropes in the sky' has been used by Scorer (1990) to distinguish rope clouds formed high in the atmosphere from those close to the earth's surface.

The significance of the level in the atmosphere where these rope clouds form is explained by figure 4.9, which shows how the temperature of the atmosphere varies with height. In the lower regions of the troposphere the normal lapse rate of temperature with height is about 6°C km^{-1}. This extends to the tropopause, whose height varies between 20 km at the equator to about 10 km in lower latitudes. It is at the tropopause that high-level gravity currents and gravity waves exist and rope clouds are sometimes formed. Above the tropopause there is a region where the temperature is roughly constant; this acts as a 'lid' to convection above which there is very little vertical mixing.

4.4.1 High-level thunderstorm outflows

Just as a rope cloud at the earth's surface defines the boundary of cold air spreading as a gravity current, a similar effect is seen as warm air spreads out at the tropopause forming the familiar anvil cloud. This was clearly shown in the photograph from an aircraft in figure 2.1. In satellite imagery anvils appear as a solid cloud mass, but occasionally distinct rope clouds separate from the cirrus cloud of the anvil and move away from the storm at the level of the tropopause.

An example of an organised tropical storm cluster which produced both a spreading gravity current at the surface and also one at high levels is shown in figure 4.10 (Kingwell, 1984). Rope clouds appeared in two separate long circular arcs; the first one L'L in the SW was associated with the spread of a cool easterly gust front. The line L'U, not exactly a continuation of the other circle, was identifiable in both visible and infra-red channels and was shown to be high cirrus cloud between 9000 and 12 000 m.

Laboratory experiments in figure 13.10 show the separation of the head of a gravity current intruding between two fluids of different density. Corresponding detached pieces of cloud are often seen at the edge of a cumulonimbus anvil intruding into the tropopause (Scorer, 1990).

Figure 4.9 Example of a temperature sounding through the atmosphere.

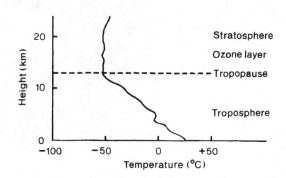

4.4.2 Rope at a warm front

The physical description of cirrus ropes at thunderstorm anvils may also
have some relevance at the leading edge of a warm front. This was thought to be
the explanation of a long rope cloud in the tropopause described by Scorer
(1990).

The most likely situation for the appearance of a rope cloud is that a warm
front with advancing cirrus should cease to be supplied with warm air. This
allows the nose at the front to advance rapidly away from the main cloud and to
become visible as a separate rope cloud. The wedge must be thin if the rope
cloud is to be detached from the main cloud. There may only be a short time
before it becomes flattened by gravity at the tropopause, between the upper
and lower air masses. Figure 4.11a shows a normal advancing warm front, based
on Bradbury (1994), and figure 4.11b shows the effect when it is retarded and
begins to recede. The break-off process is similar to that illustrated in an
experiment in figure 13.10 in which a bore separates from the head of a
gravity current moving along the interface between two fluids. The symbols to
the right of figure 4.11 are those used in weather maps and indicate the change
from the warm front in the upper figure to a cold front as the warm air ceases to
advance.

The case study shown in figure 4.12, which compares a satellite image of a
rope cloud with a corresponding synoptic chart, appears to confirm this process.
The position of the rope cloud, which extends from Iceland to the British Isles,
agrees well with the transition line between the warm and cold front marked in
the chart.

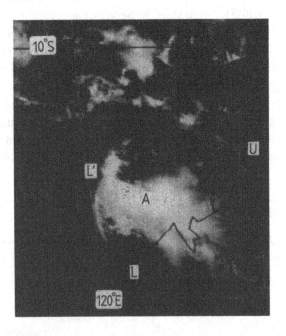

Figure 4.10 Visual satellite
image of rope clouds from a
tropical storm near Broome
at the NW coast of Australia.
The line L'L was associated
with spread of a low-level
gust front. L'U was high
cirrus cloud between 9000
and 12 000 m. A is an area of
dense stratiform cloud, with
some embedded
cumulonimbus. (From
Kingwell, 1984, with
permission from *Weather*,
Royal Meteorological Society.)

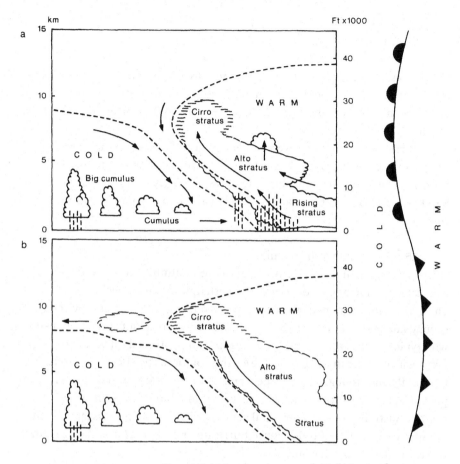

Figure 4.11 (a) An advancing warm front. (b) A warm front retarded or receding. The dashed lines show the boundaries of the cold and warm air and the arrows the direction of flow. Symbols on the right, used in weather maps, indicate the change from advancing to receding warm air.

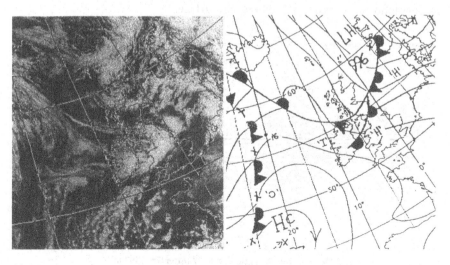

Figure 4.12 Warm front rope, between Iceland and the British Isles, 4 July 1995. The satellite image (left) and the synoptic chart (right) show the transition from a warm to a cold front.

4.5 Hurricane ropes

Hurricanes, called typhoons in the China Sea, are strong rotating storms which develop in the tropics over the ocean. The distinguishing feature of a fully developed hurricane is the circular cloud-free zone in the centre – the 'eye of the storm'.

Hurricanes form in the areas shown in figure 4.13. They occur in both the North and South tropics, but not in the equatorial zone, where the effect of the earth's rotation on atmospheric flows is small. They do not appear in the South Atlantic, where the ocean temperature is not high enough to sustain them.

4.5.1 Formation of hurricanes
Hurricanes begin in the presence of large cumulonimbus clouds, and are associated with strong vertical currents, both upwards and downwards. The downcurrents created by the rain spread out as gravity currents on the ground and at the same time the upward currents spread out at the tropopause, with typical speeds of $5-10$ m s^{-1}. As this spreading proceeds, it becomes progressively more influenced by the earth's rotation. The effect is to produce the spiral bands that give hurricanes their characteristic appearance in satellite photographs. Figure 4.14 shows the early stage of a hurricane seen off the west coast of Australia; a set of four spiral bands can be seen. A few days later a central eye appeared and the fully developed hurricane travelled as far as Madagascar, where much damage was caused.

In an active hurricane, shown schematically in figure 4.15, the eye is formed where the cloud evaporates in the downdraft at the centre. A satellite image of a fully developed hurricane, complete with eye, is shown in figure 4.16. A hurricane extracts energy from the warm sea water and usually travels until it reaches land or passes over a cooler ocean surface.

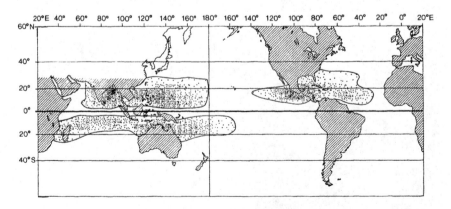

Figure 4.13 Location of initial genesis points of tropical cyclones, over the period 1958–77. (After Gray, 1979.)

Figure 4.14 The early stage of a hurricane off the west coast of Australia, 14 January 1994, showing four spiral bands.

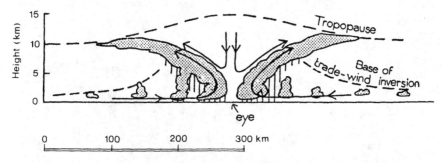

Figure 4.15 Diagram of the airflow in an active hurricane. (From Pedgley, 1961.)

Bibliography

Bradbury, T. (1994). Flying towards fronts. *Sailpane & Gliding*, (June/July), 138–41.

Gray, W.M. (1979). Hurricanes. In *Meteorology over the Tropical Ocean*, ed. D.B. Shaw, pp. 155–218, Royal Meteorological Society.

Kingwell, J. (1984). Observations of a quasi-circular squall line off north-west Australia. *Weather*, **39**, 343–6.

Matthews, D.A. (1981). Observations of a cloud arc triggered by thunderstorm outflows. *Mon. Wea. Rev.*, **109**, 2140–57.

Parmenter, F.C. (1970). Picture of the month: a 'tehuantepecer'. *Mon. Wea. Rev.*, **98**, 479.

Pedgley, D.E. (1961). *Elementary Meteorology*. Meteorological Office, HMSO.

Purdom, J.F.W. (1973). Picture of the month: meso-highs and satellite imagery. *Mon. Wea. Rev.*, **101**, 180–1.

Purdom, J.F.W. (1976). Some uses of high-resolution GOES imagery in the mesoscale forecasting of convection and its behaviour. *Mon. Wea. Rev.*, **104**, 1474–83.

Rottman, J.W. & Simpson, J.E. (1989). The formation of internal bores in the atmosphere: a laboratory model. *Q. J. R. Meteorol. Soc.*, **114**, 941–63.

Scorer, R.S. (1986). *Cloud Investigation by Satellite.* Chichester: Ellis Horwood.

Scorer, R.S. (1990). *Satellite as Microscope.* Chichester: Ellis Horwood.

Seitter, K.L. & Muench, H.S. (1985). Observations of a cold front with rope cloud. *Mon. Wea. Rev.*, **113**, 840–8.

Shapiro, M.S., Hampel, T., Rotzoll, D. & Mosher, F. (1985). The frontal hydraulic head: a micro-scale (1 km) triggering mechanism for mesoconvective weather systems. *Mon. Wea. Rev.*, **113**, 116–83.

Smith, R.K. & Reeder, M.J. (1988). On the movement and low-level structure of cold fronts. *Mon. Wea. Rev.*, **116**, 1927–44.

Smith, R.K., Reeder, M.J., Tapper, N.J. & Christie, D.R. (1995). Central Australian cold fronts. *Mon. Wea. Rev.*, **123**, 16–38.

Figure 4.16 A fully developed hurricane, showing the 'eye'.

5   Fronts and topography

The spread of a dense gas on the rotating earth will be considered in Chapter 17, and will be illustrated by some laboratory experiments on a rotating table. Coriolis forces tend to oppose the spread of a dense fluid, but a very marked effect appears when a physical barrier is present. As the fluid turns to the right (in the Northern Hemisphere) under Coriolis forces, nearly all the flow becomes concentrated into a band of dense fluid flowing parallel to a barrier. A vertical wall is not essential and a similar effect is seen with a sloping barrier.

This effect has been identified in the atmosphere in several parts of the world, and can be most marked when a large-scale cold front approaches a coastline lying from north to south. If a range of mountains lies parallel to the coast, the flow of the dense air is slowed down and a strong, intense gravity current may develop, running parallel to the coast. Such trapped gravity currents form along the coast of SE Africa and South America and off the SE coast of Australia. It should be noted that in the Southern Hemisphere the Coriolis force causes a 'turn to the left'; hence a squally wind which bursts from the south must move along the east coast.

5.1 The southerly buster

The southerly buster (or burster) is the local name given to a cold wind change which occurs in spring and summer along the coast of New South Wales in eastern Australia. The map in figure 5.1 shows the area where the southerly buster appears, alongside a line of mountains which lie parallel to the coast and reach a height of about 1500 metres (5000 feet). The squall line appears as the front of a gravity current of cold dense air which is moving from the south and is held against the mountains by the Coriolis force.

The southerly buster is a mesoscale phenomenon, developing on a time-scale of two or three days, and the details of its behaviour are not observed in detail by the regular synoptic network. The strength of the wind at a southerly buster is usually over 20 m s^{-1} and maximum gusts over 37 m s^{-1} have been recorded (Colquhoun et al., 1985). Occasionally, the leading edge of a southerly buster is marked by a spectacular roll cloud, aligned at right angles to the coast, as shown in figure 5.2. The front is not always accompanied by cloud and the abruptness of the wind change can be hazardous to low-flying aircraft and small boats. It does not usually rain at a buster, although rain often follows such a southerly change.

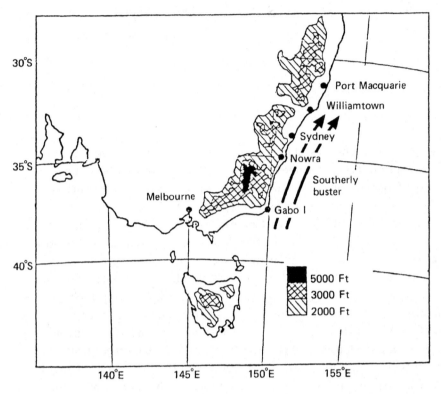

Figure 5.1 Map of SE Australia, showing the path of the southerly buster.

Figure 5.2 Roll cloud at a southerly buster, seen from near the centre of Sydney, 16 February 1985. (Courtesy of T.G. Donald.)

There can be dramatic temperature changes, with falls of 10–15°C in a few minutes in the afternoon. It has been found that the progress along the coast is strongly affected by the local temperatures and hence density difference across the front. It has been shown (Coulman *et al.*, 1985) that, incorporating this density difference and the depth of the cold air, the speeds are well predicted by gravity current theory.

Wind records made during the spring and summer show that when a front reaches the New South Wales coast it then advances northwards much faster than its penetration inland (Colquhoun *et al.*, 1985). The front does continue to move very slowly inland towards the higher ground, but its much faster steady progress along the coast in the form of a southerly buster is very clear.

The wind record made at Sydney Airport during the passage of a southerly buster is shown in figure 5.3a. At 1900 h local time the wind swung from the east to the south with a gust of maximum strength 23 m s^{-1} (about 50 mph) and blew at 10–15 m s^{-1} for about 15 minutes. It then increased and blew at over 15 m s^{-1} for 3 hours. This was one of the frequent (about 50%) occasions when pilot balloon ascents and surface wind records confirmed the presence of roll clouds in the post-frontal air, as indicated in figure 5.3b. During several of the night-time passages, although the usual pressure increase of typically 2 mb was

Figure 5.3 (a) Wind strength and direction at Sydney Airport on 11 December 1972. (b) Frontal structure in a vertical cross-section parallel to the coast. (Courtesy of J.R. Colquhoun.)

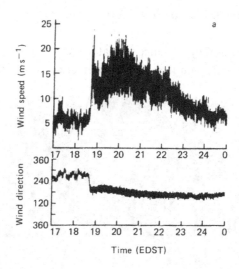

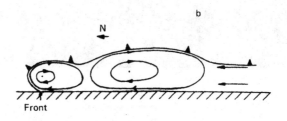

reached, there was no appreciable temperature decrease, suggesting the arrival of an internal bore on the nocturnal inversion rather than a gravity current.

5.2 Other coastal trapped gravity currents

The low-pressure systems which move around the coast of southern Africa appear to have coastally trapped features which have something in common with southerly busters (Gill, 1977).

A coastally trapped gravity current which moves from the south along the east coast of South America also seems to have much in common with the southerly buster. Between Montevideo and Rio de Janeiro there is a range of mountains about 1800 m high and a squall line called the *pampero secco* is known to move north, parallel to the coast (Georgi, 1935). Figure 5.4 shows photographs taken from a ship, in which the top view shows the extremely smooth cloud form typical of the first wave of an undular atmospheric bore. The lower two show the presence of some 'lobe' instability.

5.3 Katabatic winds and fronts

Mountains and valleys produce special winds which are much more complicated than those over flat country. Simple observations show that the local winds, close

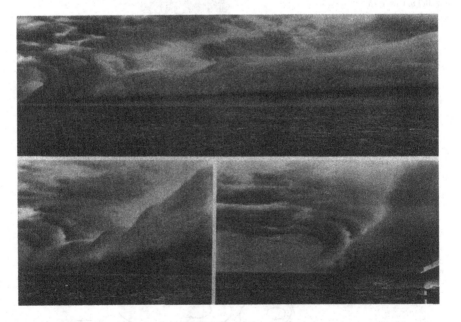

Figure 5.4 The *'pampero secco'* off the east coast of South America. (From Georgi 1935.)

to the ground, change from upslope (anabatic) flow during the day to downslope (katabatic) flow at night.

The diurnal mechanism which induces slope winds is similar to the mechanism we saw in Chapter 4 for the formation of the land–sea breeze circulation. A horizontal pressure gradient towards the slope is produced and an upslope wind results.

At night, the mechanism and the circulation are reversed, with the result that cold air drains down and away from the slope.

The upslope winds are not usually very strong, but their presence can be deduced from the formation of cumulus clouds on sunny mountain slopes. The currents which rise up slopes are used by glider pilots, but they need to fly very close to the mountain face to obtain this 'lift'. Measurements of slope upwinds have shown the maximum strength to be about 20 m from the ground, roughly equal to the wing-span of a glider.

Downslope winds may combine to produce valley winds whose influence sometimes extends out into the plains. These winds are sometimes much stronger than upslope winds, and they are capable of producing gravity currents with sharp fronts.

Some of the classic drawings of Defant (1951) are shown in figure 5.5, which illustrates the change from upslope to downslope winds. In a, towards midday, the wind up the valley has set in; it feeds the upslope winds and is supported by

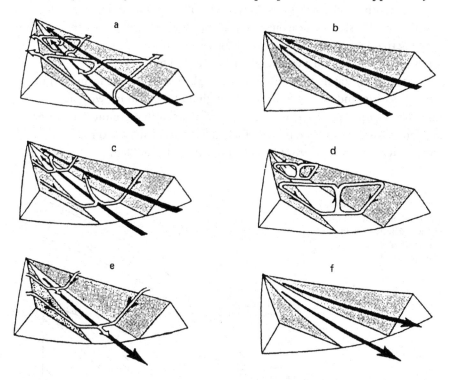

Figure 5.5 The development of down-valley winds.
(After Defant, 1951.)

the return flow from the greater heights in the centre of the valley. In the late afternoon, b, the upslope winds have ceased to blow, but the valley wind continues for a short time. In c the downslope winds have started to blow, but since the valley wind is still blowing upslope, no cold flow down the valley results. In d the downslope winds alone remain, and in e the cold downslope air overcomes the up-valley flow and moves forwards, sometimes as a sharp cold gravity current front. This onset of a katabatic wind as a sharp gust front is a common feature and at some places this onset has been described as an 'air avalanche'.

In some recent observations a different series of events was seen (Thompson, 1984). The temperature changes in the valley floor seem to have initiated the katabatic wind, which moved down the valley floor as a gravity current front. Here, the katabatic wind was more likely to have been a result of the accumulation of cold air in the valley, which by reason of its greater density moved down the valley under gravity. Figure 5.6 shows the situation at the mouth of the Red Butte Canyon in the early evening.

The combination of several valley winds flowing down from a mountain range may form a strong dense gravity current, resembling a cold outflow from a thunderstorm. In two cases of night-time cold air drainage down the eastern slope of the Rocky Mountains in Colorado the cold air progressed as a gravity current to the city of Boulder, 25 km distant (Blumen, 1984). On both nights the drainage front occurred abruptly as a gravity current moving at about 3 m s^{-1}, the total depth of which was less than the 300 m height of an instrumented tower. Figure 5.7 shows the acoustic sounder (sodar) trace on one of these nights, showing the sharp front and a series of pulsations, resembling those seen in atmospheric bores, with a wavelength of about 10 km.

Very strong katabatic winds, known as 'glacier winds', are formed from large ice sheets, and mostly continue throughout the 24 hours, with no interrupting upslope phase. The strongest of all are found around Antarctica, and figure 5.8 shows a wind record of a katabatic wind at Mawson, Antarctica, which after a strong gust front reaches a strength of over 20 m s^{-1}.

Figure 5.6 The initial stage of katabatic flow, a pressure gradient flow at the mouth of the Red Butte Canyon, Arizona, USA. (After Thompson, 1984.)

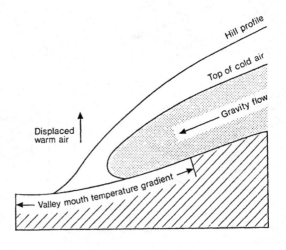

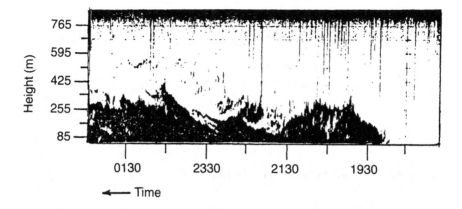

Figure 5.7 Acoustic sounder trace showing the arrival of drainage flow at Boulder, Colorado, on 8 October 1980. (Courtesy of W. Blumen.)

Figure 5.8 Wind speed and direction at onset of typical katabatic wind, 21 August 1960, Mawson, Antarctica. (After Streten, 1963, courtesy of Academic Press.)

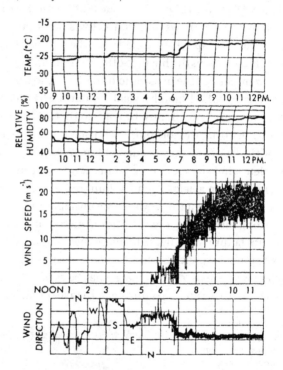

Bibliography

Blumen, W. (1984). An observational study of instability and turbulence in nighttime drainage flows. *Boundary Layer Meteorol.*, **28**, 245–69.

Colquhoun, J.R., *et al.* (1985). The southerly burster of eastern Australia. *Mon. Wea. Rev.*, **113**, 2090–107.

Coulman, C.E., *et al.* (1985). Orographically forced cold fronts – mean structure and motion. *Boundary Layer Meteorol.*, **32**, 57–83.

Defant, F. (1951). Local winds. In *Compendium of Meteorology*, pp. 665–72. American Meteorological Society.

Georgi, J. (1935). Pampero secco von 17 July 1935 (in German). *Der Seewart*, **7**, 199–205.

Gill, A.E. (1977). Coastally trapped waves in the atmosphere. *Q. J. R. Meteorol. Soc.*, **103**, 431–40.

Streten, N.A. (1963). Some observations of Antarctic katabatic winds. *Austr. Meteorol. Mag.*, **42**, 1–23.

Thompson, B.W. (1984). Small-scale katabatics and cold hollows. *Weather*, **41**, 15–53.

6   Environmental problems: atmosphere

6.1 Gravity current hazards to aircraft

Violent shifts in wind speed and direction, known as 'wind-shear' and turbulence related to thunderstorms are the principal meteorological hazards to aviation today. Although a well-developed thunderstorm is usually identifiable and can be avoided by planning the course of the flight, serious problems are associated with the cold air outflows from convective storms. Sudden changes in horizontal and vertical wind speed associated with the gust front of a cold outflow can be especially dangerous during take-off and landing.

A reminder of the hazards to be found at the front of a gravity current is given in figure 6.1. Although a shadowy outline has been sketched to show the form of the gust front, it must be realised that, unlike some other spectacular parts of the storm, the cold outflow is usually completely invisible. The leading edge of the front of a gravity current is maintained as a very sharp divide, so that a complete wind change may occur in a few tens of metres. As we follow this sharp interface to a height of a few hundred metres, it begins to roll up, forming billows. These grow until they become unstable and then decay, filling a layer with strong turbulence several hundred metres deep. Aircraft have undergone structural failure when flying through the billow area, but the greatest hazard of flying through a single gust front is the almost instantaneous change in air speed which may be imposed. This change may be as much as 60 knots (30 m s^{-1}) in a few seconds; there are times during both take-off and landing when the aircraft

Figure 6.1 An invisible dense outflow approaching Dulles Airport, near Washington, DC, through a detection system of pressure sensors. (Courtesy of the American Meteorological Society.)

will not have this reserve above its stalling speed, with serious consequences.

A particularly difficult situation can occur when an aircraft, engaged in take-off or landing, flies right through a downburst cell in which the descending cold air is spreading out in all directions (Fujita, 1981). Figure 6.2 shows events which can happen when an aircraft flies through both sides of a downburst during a landing or take-off. The aircraft making the approach in figure 6.2a is flying at the correct airspeed and glide path when it meets a sudden headwind caused by the first part of the downburst; this results in an increase of airspeed and a climb above the correct glide path. The pilot then reduces his airspeed and also steepens his approach. These procedures combine to make it impossible for him to cope with the tailwind and downcurrent which he encounters in the far side of the downburst. As shown in figure 6.2b the hazards are very similar during a take-off through a downburst.

It is possible to reduce the dangers of flying through such cold outflows by warning the pilot of the presence of a downburst, but there remains the very serious problem of what have become known as 'microbursts'.

6.1.1 Microbursts

A microburst, as its name suggests, is a very small downburst of cold air from a thunderstorm. It is by no means small in the strength of the wind-shear and downcurrent it can produce, but its shortness in duration, perhaps only two to four minutes, makes it very difficult to detect in advance. Microbursts have been

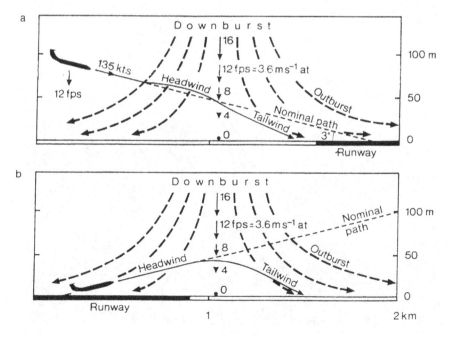

Figure 6.2 The effect of the airflow in a downburst on an aircraft (a) approaching an airfield, and (b) taking off from an airfield. (After Fujita & Caracena, 1977.)

studied by Professor Fujita of Chicago University, who has recorded 18 aircraft accidents round the world which have been attributed solely to this phenomenon (Fujita, 1985).

Measurements using doppler radar (Wakimoto, 1982) and photographs of dust clouds and patterns in damaged crops and trees have lead to the model shown in figure 6.3. The observations show that an outflow microburst is surrounded by a ring vortex, spreading along the ground. This is exactly as seen in the very early stages of laboratory experiments in which a quantity of dense fluid is allowed to descend and then spread out along the ground. These experiments and the explanation of the formation of the vortex will be described in Chapter 12.

The descending air which forms a microburst is carried down by rain, and its density increases as it is cooled by the evaporation of water drops. If the cloud-base is low the falling rain will not all disappear and the microburst will be 'wet'. In a 'dry' microburst, all the rain has evaporated during the descent from a very high cloud-base and an intense microburst is formed.

When the descending microburst reaches the ground the form of the strong vortex at the leading edge of the outflow is sometimes made visible by dust raised in the strong wind, as shown in figure 6.4.

6.1.2 Detection of wind-shear

Figure 6.3 The ring vortex believed to form at the edge of an outflow microburst.

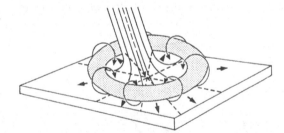

Figure 6.4 The leading edge of a microburst showing the vortex outlined by a dust cloud. (Courtesy of Brian Waranauskas.)

The hazards of wind-shear to aircraft are now appreciated; according to the International Civil Aviation Authority sudden shifts in the wind have killed more air passengers in the USA than any other cause.

The first generation of wind-shear detectors consisted of a number of wind recorders and pressure sensors placed around an airport to monitor sudden changes of wind and pressure. An early example known as Low Level Windshear Alert System (LLWAS) was set up at Dulles Airport (see figure 6.1) and was successfully used to detect gust fronts using the associated pressure jumps (Bedard *et al.*, 1977). Similar schemes operate at many airfields in the USA; they can give useful information to pilots on the approach to landing but cannot deal with a sudden microburst as it forms. For example, at the accident at Charlotte in 1995 due to an unusually severe microburst the LLWAS only sounded the alarm 22 seconds after the aircraft had struck the ground.

The new generation of wind-shear detectors attempts to measure the movements of the air above the airport and its surroundings. Doppler radar using reflection from airborne objects, such as raindrops, can detect whether they are moving towards or away from the transmitter. The difference from ordinary weather radar is that in this case scanning is needed close to the ground, needing a very narrow beam width. Some aircraft already carry airborne wind-shear systems; these measure the change in speed of airflow over the wing but give a microburst warning which is too late. More effective portable doppler radar detectors are now available for mounting in aircraft and are increasingly being used.

All these systems can only spot microbursts in action; what is really needed is to predict the formation of a microburst. This is now being achieved by developing computer software to combine airborne measurements of temperature and humidity with weather radar data to detect potential microbursts.

Now the significance of microbursts is understood, training in how to cope with them is being given to pilots on simulators. Combined with the detection techniques described above, the hazards of wind-shear are being steadily overcome.

6.2 Spread and dilution of a dense gas

The number of accidental releases of dangerous heavy gases grew during the 1980s and 1990s. One reason was the increased distribution and use of potentially dangerous liquid natural gas (LNG).

LNG was first developed in the USA in the 1930s and its use increased until an explosion in 1944 at Cleveland. Tanks containing 4200 m^3 of LNG split and the cold vapour spread into streets, ignited and engulfed houses, causing the deaths of 150 people.

After that, no new plants were built until the late 1950s when new safer tanks had been developed. Large-scale sea transport of LNG (up to 125 000 m^3)

and its storage in terminals adjacent to centres of population gave rise to concern about the consequences of an accident. Fortunately, these fears have not been realised to date.

Nevertheless, a number of serious accidents have happened in which containers of liquid fuel gases were ruptured for various reasons and dense gas was released. The liquid and rapidly evaporating gas spread as a gravity current into the surroundings and a serious explosion or fire took place. Other events have involved highly toxic gases which almost invariably result in the formation of a dense, low-lying plume. The following list records the more serious events, involving either flammable or toxic gases.

1959. Meldrim, Georgia – A liquified propylene gas (LPG) tank on a train ruptured and a cloud of gas spread over a picnic ground before igniting; 23 people died.

1969. Crete, Nebraska – A rail tank containing 72 tonnes of liquified ammonia was completely ruptured in an impact following derailment; nine people were killed.

1973. Potchefstroom, South Africa – An ammonia storage tank failed, releasing 38 tonnes of gas; 18 people died.

1978. Los Alfraques, Spain – A road tanker carrying 43 m^3 of LPG sprang a leak as it passed a camp site. The vapour spread over the camp and burst into flame; 150 people died.

1978. A highway near Mexico City – A road tanker spilled LPG after a collision; 20 people died.

1981. Montanas, Mexico – A train derailment resulted in the fracture of tank cars containing chlorine. Over 50 tonnes of gas were released in a few minutes; 17 people died and 1000 needed hospital treatment.

1984. Mexico City – Explosion at a distribution centre for LPG adjacent to a shanty town; over 500 people died.

1984. Bhopal, India – 40 tonnes of methyl isocyanate were released from a storage tank. The plume passed over a shanty town; about 2500 people died.

The need for the assessment of the hazards associated with such accidental releases of dangerous heavy (denser than air) gases has led to the development of a large number of simplified models of dense gas dispersion. The physics of some of the processes is not well understood and their description in the models involves a good deal of empirical input.

6.2.1 Dense gas dispersal models

Different types of mathematical model have been developed for the description of dense gases (Blackmore *et al.*, 1982; Britter & McQuaid, 1989); they can be separated into three categories (A) fully three-dimensional, (B) depth averaged, and (C) box models.

The three-dimensional models attempt to solve suitable approximations of the partial differential equations for conservation of mass, momentum and species to solve the mean velocities and mass concentrations. The main differences between the models available are not in the basic equations employed but in the numerical methods used to solve them. These models have the advantage that they can simulate a variety of release conditions, terrain and other obstacles present in the flow. They are expensive to run and uncertainties remain about the way the turbulence is modelled, justifying a simpler treatment.

Depth-averaged models further simplify the equations for the mean velocity and concentration by averaging vertically over the depth of the gas cloud. The turbulent mixing is then incorporated into the model as the rate at which the cloud height increases due to turbulent mixing. The main feature of these models is the treatment of the turbulent entrainment velocities. For example, in one model (Ermak *et al.*, 1982) it is assumed that the turbulent entrainment can occur at both the top and the edges of the cloud. One of the advantages over the fully three-dimensional group is the reduced cost of computation, but they maintain flexibility in computing cross-stream profiles.

'Box' models are far the most popular of all the types of model of dense gas dispersion (van Ulden, 1974; Fay & Ranck, 1983) and general reviews have described in detail the different forms (Wheatley & Webber, 1984). Box models make the greatly simplifying assumption that the profiles of mean velocity and concentration in the cloud are known. The simplest box models assume that the cloud has the shape of a flat circular cylinder, with uniform height and radius, in which the gas concentration is uniform. Some models assume different box profiles from the rectangular shape, but there is little advantage in this complication.

The schematic figure 6.5 shows a box model. The cloud is released at $t=0$, with its centre at the point marked O, and is assumed to maintain a cylindrical shape, with height $h(t)$, radius $r(t)$ and uniform density $\rho(t)$ as it is advected downstream.

The equations covering the evolution of radius r, volume v and distance to the cloud, x are:

$$v = \text{area} \times \text{height} = \pi r^2 h$$
$$dr/dt = \text{front speed} = c(g'h)^{\frac{1}{2}}$$
$$dv/dt = \pi r^2 u_z + 2\pi r h_r = \text{top entrainment} + \text{front entrainment}$$

$$dx/dt = u(z) = \text{wind speed at height } z, \text{ whose specification varies from one}$$

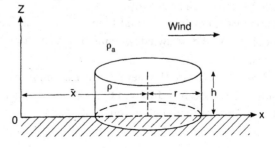

Figure 6.5 Sketch of a box model of a spreading cloud of dense gas. The centre of the cloud is at the point O when $t=0$. The cloud maintains a cylindrical shape, with radius r, height h and density ρ as it is advected a distance x downwind.

model to another.

This model is only appropriate when the bouyancy forces are significant. Allowance has to be made as the shape is reached when atmospheric turbulence becomes the dominant mechanism for the dispersion of the cloud. In all the models, the specification of the entrainment velocities remains the chief question to be resolved.

To complement the theoretical modelling efforts, a number of laboratory and field experiments has been carried out. The relevant laboratory experiments will be described in detail in Chapters 12, 13 and 14. Two of the largest field trials performed in Britain have been at Porton Down and, more recently, at Thorney Island.

6.2.2 Porton Down and Thorney Island field experiments

These large-scale experiments were designed to provide data which could be used for the following purposes:

1.   To enable predictive models to be verified.
2.   To further the understanding of the physical processes involved and to test model hypotheses.
3.   To provide information on the scaling laws so that better uses could be made of small-scale work in wind tunnels and water channels.

In the field experiments at Porton Down clouds of dense gas, 40 m^3 in volume, were released suddenly into the atmosphere with a minimum of disturbance (Picknett, 1981). The clouds had densities in the range 1.03 to 4.2 relative to air, and were made visible by coloured smoke. The required density was obtained by mixing air and Refrigerant 12 in suitable proportions.

The gas was contained in a tent made of plastic sheet in the form of a cube of side about 3.5 m. A photograph of an early form of the tent, ready for release, is shown in figure 6.6. The roof, supported by four light alloy pillars, remained in position after the collapse of the walls.

Photographs were taken of each gas cloud from two directions after release to give a view in plan and elevation, and gas concentration measurements were made.

Each experiment at Thorney Island (McQuaid, 1985) was begun by releasing the much larger volume of 2000 m^3 of a heavy gas nearly instantaneously from a cylindrical container into an ambient wind. The heavy gas was a mixture of nitrogen, Refrigerant 12 and smoke. In most of the experiments the mixture was adjusted so that it had a density about twice that of the surrounding air. The container was 14 m diameter and 13 m high and was made of plastic sheeting whose sides could be drawn down quickly to the ground by elastic cords in about two seconds.

A second series of Thorney Island experiments was carried out in which the spreading dense gas met different types of obstacle. These obstacles were a 5 m high solid fence, a 10 m high porous fence and a cube with sides 9 m high.

6.2.3 Release into calm air

Experiment 8 in the Porton series consisted of a release in the evening of a day when conditions were almost windless. The cloud collapsed symmetrically to the ground, and the vorticity generated at the cloud edge as it collapsed was concentrated at the front edge, forming a radially expanding vortex ring. It was apparent that at this stage most of the gas was contained within this 'raised rim'. Figure 6.7 is a photograph of this effect, seen from above.

6.2.4 Release into wind

After the initial collapse, the motion is mainly horizontal, as the front advances with the kind of gravity current head described in earlier chapters. The majority of the mixing with the ambient fluid occurs at the front of the current, and the density profile behind the head consists of a layer of roughly uniform density gradient, lying above a layer with uniform density.

However, in an ambient wind the structures of the upwind and downwind fronts are quite different from each other. Figure 6.8 shows a release at Thorney Island in a wind of 4.7 m s^{-1}; the wind in this view is blowing from left to right and the steep front upwind can be compared with the much flatter one on the downwind edge. In figure 6.9 the speeds of the upwind and downwind fronts from the trials at Thorney Island are compared with the empirical formula (Simpson & Britter, 1980) to be derived later in section 11.5. Stability considerations suggest that the upwind front is more stable than the downwind front, so more mixing would be expected at the downwind front.

During the gravity-spreading phase the stably stratified upper layer tends

Figure 6.6 Early form of a container for releasing a large volume of dense gas at Porton Down. (Courtesy of Health & Safety Executive.)

to suppress any turbulent mixing with the ambient fluid. However, in a turbulent environment there will be mixing due to the impinging external turbulence, and as the density driving effects diminish, the external turbulence takes over as the main control. As viewed experimentally this problem will be dealt with in Chapter 14.

Figure 6.7 Release of dense gas into calm air seen from above in the Thorney Island experiments. (Courtesy of Health & Safety Executive.)

Figure 6.8 Profile and plan views of a dense cloud at various times after release in a 4.7 m s⁻¹ wind in a field experiment at Thorney Island. (Courtesy of R.G. Picknett.)

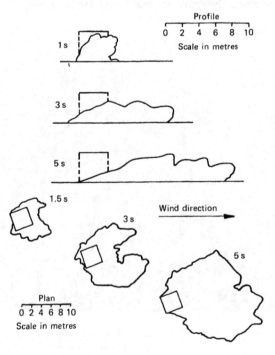

6.2.5 Maplin Sands experiments

The Maplin Sands experiments were performed in the summer of 1980 (Puttock *et al.*, 1984). As in the two series of experiments described above, the aim was to study the dispersion of dense gases, but there were important differences in the procedure.

A total of 34 spills of liquified gas onto the sea surface was performed. The gases used were refrigerated liquid petroleum gas (LPG) and liquid natural gas (LNG), in quantities up to about 20 m³. An important feature of these experiments is the very large cooling of the air into which the liquid evaporates due to the latent heat required to cause the evaporation. In the later stages the main difference between the behaviour of the LPG and LNG spills is that LPG remains denser than air, even if warmed up to ambient temperature. The behaviour of LNG is more complicated, as the gas is initially dense due to its low temperature ($-162\,°C$), but heat transfer from the water and air will ultimately render the gas positively buoyant.

Release of the liquid was either continuous or instantaneous. In continuous spills, liquid was released from the end of a tube near the water; for instantaneous spills the liquid was poured into a floating, open-topped barge. When this was submerged the water flowed in, displacing the liquified gas.

The site was a flat area of tidal sands on the north side of the Thames estuary and the releases were made 350 m from the shore at high tide during periods of offshore wind. An array of 360 instruments enabled measurements to be made of gas concentration, temperature and wind.

In one of the LNG spills, no. 29, the supply was kept steady at 4.1 m³ min.$^{-1}$ for 225 s into a steady wind of 7.4 m s^{-1}. The photograph in figure 6.10a shows the fully developed plume, extending over a length of 250 m. The results of all the spills were compared with the predictions from the HEGADAS II model which was developed by Shell Research. Figure 6.10b shows a plan view of the

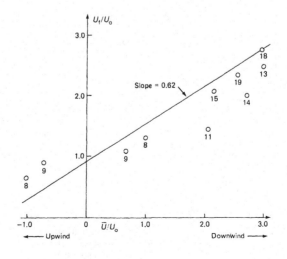

Figure 6.9 Plot of the upwind and downwind front speeds as functions of the mean wind speed for the Thorney Island releases. The solid line represents the empirical line developed by Simpson & Britter (1980).

spill, together with the limits of the visible plume predicted by the HEGADAS model, showing a close agreement.

6.3 Gravity currents of gases in mines

Extensive research has been done on the stably stratified layer of a contaminated gas which can form along the roof (or floor) in the ventilated roadways of coal

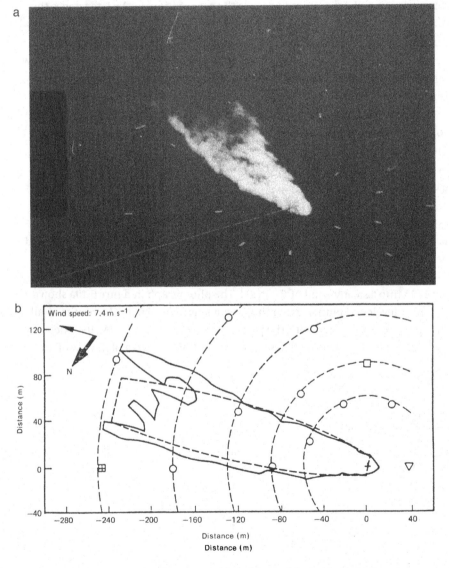

Figure 6.10 Release of dense gas in the Maplin Sands experiment. The dashed line in (b) shows the limits predicted by the HEGADAS 11. The symbols show the positions of the various sensors. (Courtesy of J. Puttock.)

mines (Fletcher *et al.*, 1973). In fact, some of the earliest research on the behaviour of gravity currents, done in 1942 by Georgeson, was based on the free streaming of such a buoyant gas in a sloping channel.

Stably stratified layers have arisen in mine roadways when methane, which is lighter than air, is released from a source near the roof. If the stable density gradient of the gas becomes sufficiently large, mixing by turbulence will no longer be sufficient to disperse the gas. A layer can form and persist for long distances; if the ventilation is downhill a gravity current front will form. This front may move backwards, or become arrested, or it can even flow up the slope against the ventilation, as illustrated in figure 6.11. In cases such as this, it would be necessary to know the ventilation flow rate which would ensure safe approach from the upwind side.

A large number of other environmental and industrial processes involve turbulent mixing in stratified flows – the important non-dimensional number concerned with this type of mixing is the Richardson number. This relates the restoring force on a fluid when it is displaced from the equilibrium position to the forces available to mix the fluid.

To establish an overall Richardson number, Ri_0, consider a layer of fluid with depth H, mixing between two horizontal planes, one above the other, of average density ρ, and total density difference $\Delta\rho$. Suppose the fluid of the lower plane moves horizontally with velocity ΔU relative to that of the upper boundary, and that the vertical gradients of density, $\Delta\rho/H$, are constant. If we move a fluid parcel a small distance, d, upwards then the change in potential energy is $g\Delta\rho d/2H^2$, and the available kinetic energy is $\rho(\Delta Ud)^2/2H^2$. The ratio of potential energy change to kinetic energy change is then

$$Ri_0 = g\Delta\rho H/\rho(\Delta U)^2.$$

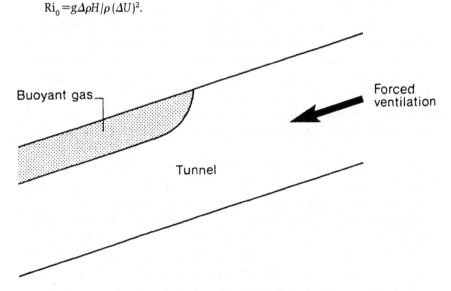

Figure 6.11 The effect of forced ventilation on the front of a buoyant gas on the roof of a tunnel in a mine.

If $Ri_0 > 1$ more potential energy is required to mix the fluid in this way than is available from the kinetic energy in the velocity shear. Further theory, supported by experimental results, shows that $Ri_0 > \frac{1}{4}$ is a necessary condition for stability.

The overall Richardson number, Ri_0, is also called the 'average gradient Richardson number'. In the atmosphere and in the oceans, Ri_0 is of the order of 10 or more, but, nevertheless, turbulence and mixing often occur. In such cases detailed measurements show that the local 'gradient Richardson number' has values near $\frac{1}{4}$.

Work on the behaviour of roof layers in mines has extended over many years at the UK Safety in Mines Research Establishment, now the Health and Safety Executive. It has been established that the behaviour of a roof layer in a given roadway is governed by a fundamental parameter called the layering number. The inverse cube of the number is proportional to the Richardson number and it has been shown (Bakke & Leach, 1962) that a layering number greater than 3 is required in a downhill ventilation roadway of small inclination to prevent a layer from backing uphill against the ventilation. Moreover, experiments (Bakke et al., 1964) have shown that if the layering number is much less than 3 a layer can back up a great distance against the ventilation.

6.3.1 Explosions caused by a layer or accumulation of methane

Two examples give good illustrations of the behaviour of a buoyant gas in a complex situation.

(a) Cambrian Colliery

In 1965, 31 men were killed in an underground explosion at Cambrian Colliery. During the following investigations an opportunity occurred to observe another layer formed by reproducing the ventilation conditions that were believed to have existed at the time of the explosion (Leach & Thompson 1968).

Concentration measurements of methane were made at a series of points in the return ventilation roadway, which carried air from the mine-face. This roadway was 120 metres long and sloped down at 1 in 33 from the face in the direction of the ventilation flow. It was believed that methane was entering the roadway from the roof between 90 and 110 metres from the junction with the face.

The ventilation rate at the time of the accident was believed to have been either 0.15 or 0.1 m s^{-1}, and measurements were made with each of these flows, taking a time of half an hour for conditions to stabilise. The form of the methane observed near the roof under these two different conditions is shown in figure 6.12. It can be seen that in the 0.15 m s^{-1} case the methane is contained by the ventilation flow, but in the case of the weaker ventilation much of the methane reaches the corner at the end of the mine-face. The methane distribution was also observed at a ventilation velocity of 0.38 m s^{-1}, when the methane layer dispersed in a few minutes.

Smoke was injected into the methane flow in both the low-speed cases and it was made clear that the methane gas flow was backing against the ventilation. The layer number was found to be 0.7, much less than the critical number of 3. At such low layering numbers buoyancy forces predominate and a layer can rapidly back against the ventilation flow.

It was calculated that there was a volume of about 24 m³ of methane in the layer at the slower ventilation velocity, which is sufficient to explain the observed velocity of the explosion.

(b) *Golborne Colliery*

In 1979 an incident occurred at Golborne Colliery in which a large accumulation of methane was concentrated (Mercer, 1981). An idealised diagram of the heading in figure 6.13 shows the accumulation of methane in the upper part; this had occurred after a breakdown of the auxiliary ventilation. When an auxiliary fan was restarted the airflow caused the methane to be released at the mouth of the tunnel. Some of this gas was ignited and an explosion took place, killing 10 workers.

One question that was raised was whether, on turning on the fan, the accumulation of methane would move 'en masse' or would be eroded away more slowly by air channelling beneath it. A laboratory experiment was carried out as illustrated in figure 6.14. This was performed, as are many described in the second part of this book, by an 'inverted' experiment in which a dense salt solution took the place of the methane gas, and fresh water was pumped through two lengths of 50 mm glass tube.

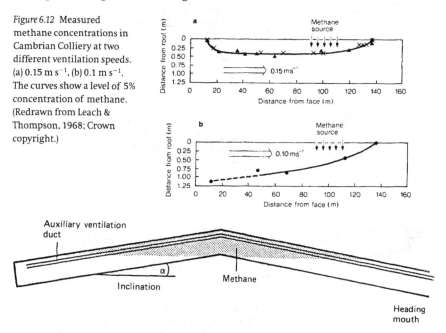

Figure 6.12 Measured methane concentrations in Cambrian Colliery at two different ventilation speeds. (a) 0.15 m s⁻¹, (b) 0.1 m s⁻¹. The curves show a level of 5% concentration of methane. (Redrawn from Leach & Thompson, 1968; Crown copyright.)

Figure 6.13 Idealised geometry of the heading in Golborne Colliery. (From Mercer, 1981; Crown copyright.)

All the tests were carried out with the interface between the brine and water initially as shown in figure 6.14a. When water was introduced at the 'face end' the brine accumulation moved down the tube as a plug. There was no tendency for the water to channel over the brine.

The plug motion continued until the leading edge of the interface reached the point where the tube changed from its downward slope to being upwards. The water then began to channel its way over the brine, forming a wedge of water above it, as shown in figure 6.14b. The rate at which the wedge penetrated into the brine was about twice that measured in the earlier stages, suggesting that about half the cross-section of the tube was occupied by the water. Before water reached the outflow end of the tube, brine was pushed out at the same rate as the water entered the apparatus at the other end.

So, 're-inverting' the geometry, it was concluded that when methane had 'locked off' the heading, the accumulation would move as a plug once the fan was started. The rate at which the methane would leave the heading mouth would equal the rate of the air input from the fan. This would apply, both before and after the air had broken through the crest, until the wedge of air reached the heading mouth. After that, a layer of methane would remain in the downward sloping leg of the heading, and would be slowly eroded.

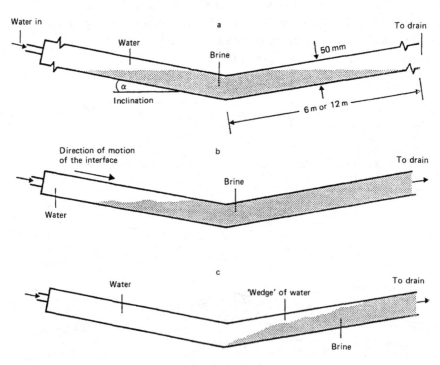

Figure 6.14 Laboratory model of the Golborne Colliery incident. (a) Layout of model. (b) Motion of brine as a plug. (c) Water channelling over the brine once the dip has been cleared. (From Mercer, 1981; Crown copyright.)

This type of work has been extended to include the study of buoyancy-driven exchange flows in damaged ducts, a flow which might arise in the coolant ducts of nuclear reactors. Some of the phenomena investigated in coal mines may also occur in vehicular tunnels and in underground garages.

6.4 House ventilation flows

Very few houses in the world have artificially driven and controlled ventilation systems. Most rely on mixing by convection currents produced by heating from internal sources, and by open windows and doors. These open windows and doors can be responsible for excessive losses of heat in cold countries and invasion of hot air in hot climates, as gravity currents of dense or buoyant air pass through these spaces (Brown, 1962; Shaw & Whyte, 1974).

Flows through open doors can account for a significant part of total heat losses in housing. They can also determine the distribution of indoor contaminants within a building. Even at quite small temperature differences (a few degrees) between the exterior and interior air, buoyancy forces are significant and a gravity current flow is established through an open doorway. This flow may cause a loss of heat due to the intrusion of cold air along the floor, or a heat gain in a refrigerated room, in which case the incoming air flows along the ceiling.

6.4.1 A model of a two-dimensional flow

Small-scale laboratory experiments can illustrate the kinds of flows which occur in houses and enable calculations to be made about the full-size flows. In small-scale experiments, instead of using hot and cold air, water is used as the working fluid and density differences are produced by dissolving varying amounts of salt. There are three advantages of this technique:

1 Water allows for striking flow-visualisation using shadowgraph imagery or dyes to mark different flows.

2 Measurement of flow velocities may be made directly and equivalent temperature measurements can be made by measuring the density of the salt solution within the model.

3 The correct range of the relevant dimensionless numbers concerned with viscosity and diffusion can be maintained, enabling accurate quantitative measurements to be made.

The flow through an open doorway at the end of a passage is modelled in the experiment illustrated in figure 6.15. When the door is opened the dyed dense fluid enters the building as a gravity current which fills half the depth of the door and advances at a uniform velocity along the floor. Outside the house at the same time the less dense fluid can be seen rising up the outer wall and mixing with the surroundings.

When it reaches the end of the passage the gravity current is reflected and can be seen travelling back towards the door. The dense fluid almost fills the passage, but there is a small space above filled with lighter fluid which appears to have difficulty escaping.

Some results of such experiments are shown in figure 6.16, where the distance covered by a series of gravity currents of cold dense air entering a model house door is plotted against time, on log–log scales. Data from experiments of different size and density difference have been non-dimensionalised and all fall on the same straight line of gradient 1. This shows that each current advanced at a constant speed during its journey from the door to the end of the corridor. The arrow points to an experimental point measured after opening the front door of the author's house; this lies on the same line as that determined by the small-scale experiments. The working substance here was, of course, air and not water as in the smaller-scale results; a temperature difference of 3 °C in air corresponds to about a 1% density difference in water.

6.4.2 Design of buildings with gravity-driven ventilation

Some ideas for the design of buildings in which gravity-driven ventilation is employed can be drawn from Nature, and in particular from the termites.

Details of a termite mound and its ventilation system are shown in figure 6.17. The purpose of the structure visible above ground, a wedge-shaped tower over 3 m high, is to collect heat from the sun's radiation and to maintain a rising

Figure 6.15 Laboratory model of dense flow entering a house. A gravity current of dense fluid flows through an open door into a passage between parallel walls.

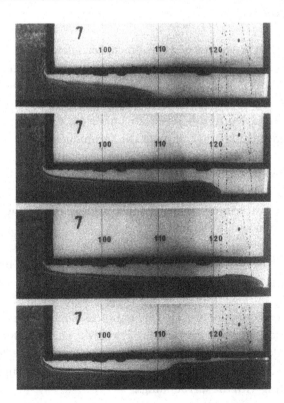

current of air through the whole building. Additional airflow is generated by the wind blowing across the top of the tower. Most of the living space is below ground level and is kept at a steady temperature by the termites opening and closing ventilation channels. The system is very effective; measurements in Australian surroundings varying from 3 °C to 40 °C have shown that the temperature within the structure is kept within 1 °C of a steady 30 °C.

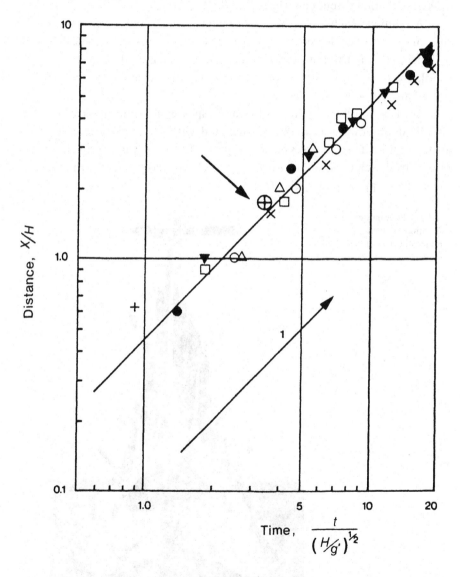

Figure 6.16 Distance X travelled by experimental gravity currents along a passage in time t (on log–log scales). X is non-dimensionalised by the height H and time t by $(H/g')^{\frac{1}{2}}$. Points marked $\times$, $\bigcirc$, $\square$, $\blacktriangledown$, $\triangle$, $+$ refer to laboratory experiments of different density. $\oplus$ (marked by arrow) = experimental point measured after opening the front door.

Atria, which form part of a number of modern buildings, present challenges in the design of efficient ventilation systems. The large glazed areas associated with atria have high solar gains in summer and large heat losses in winter and present problems in the air movement and temperature distribution over numerous levels of living space. However, by suitable design, the hot air rising in an atrium can be used, as in the termite mound, to operate gravity-driven ventilation throughout the rest of the building.

A building which was designed with reference to laboratory experiments is the University of Seville Department of Humanities, shown in figure 6.18 (Baker & Linden, 1991). In this four-storey building the atrium is used to drive ventilation of the surrounding rooms, while at the same time retaining a cool pool of air at ground level.

The laboratory model, made of clear perspex, enabled the flows to be easily visualised. Water was used as the working fluid, with sources of heat within the atrium represented by sources of salt solution. The model was inverted in a large tank of fresh water and viewed through an inverted video camera so the dense salt solution appeared to 'rise' in the model.

Figure 6.17 Buoyancy-driven ventilation in a termite mound. (Courtesy of *New Scientist*.)

The solar-heated upper atrium was found to be sufficient to drive flow at the required rate through the lecture rooms. The pool of cool air at the base of the atrium was eroded by the plume of air emerging from the rooms. Buoyant air from occupied rooms moved upwards with little disturbance of the pool but air emerging from cooler unoccupied rooms was more destructive. Disturbance of the cool pool caused by openings to the atrium at ground level suggested that main circulation routes from outside should be on the first floor rather than at the ground.

A remarkable building using these and other novel ideas was build at Cambridge in 1994. The Ionica building uses the 'stack effect' of rising air in a glass-roofed atrium extending over three storeys, supplemented by additional airflow derived from six wind towers mounted on the roof.

Similar tests continue on other buildings and are expected to provide useful information on atrium design.

6.4.3 Fires in buildings

Most of the heat from a fire in the open air is lost to the atmosphere and the fire spreads at a steady rate. A fire inside a house is very different; its development depends on the geometry of the room, the height of the ceiling, the position of doors and windows and the presence of staircases.

Within a minute of the start of a fire near the ground the burning material gives off smoke and heat which rise to the ceiling. Figure 6.19 is a calculated temperature plot of such a rising hot plume in a room with height-to-width ratio of $\frac{1}{2}$. The front of a hot gravity current, seen moving along the ceiling to the right, forms a layer on the ceiling which becomes thicker and hotter as the fire continues. After a few minutes, due to the radiant heat from this layer, everything in the room bursts into flame. The rest of the house will become

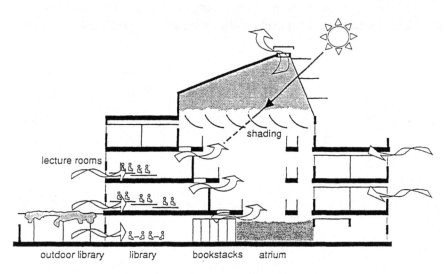

Figure 6.18 Simplified section of the University of Seville Department of Humanities.

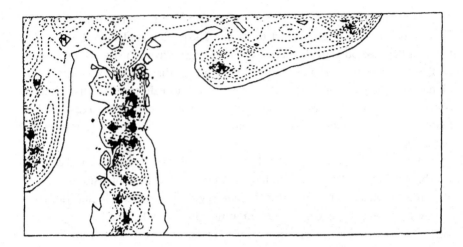

Figure 6.19 A sample calculated temperature plot of flow along the ceiling in a fire, showing a rising hot plume and a gravity current front to the right.

rapidly affected as smoke and toxic gases spread through doorways into other rooms until the whole house is on fire.

Understanding how fires spread is important for everyone: serious fire accidents have happened in large buildings when people did not understand what was happening and were unable to take appropriate action. At the Bradford football round in 1985 the fire started when litter beneath wooden seating caught fire and the space beneath the stand acted like a room in a house. Gaps in the woodwork allowed smoke and flames to rise to the roof of the stand and form a layer which radiated heat on the people beneath, many of whom were killed before they realised what was happening.

Another serious accident occurred at King's Cross tube station in 1987 where a fire formed in litter spread rapidly up the slope of an escalator.

Bibliography

Baker, N. & Linden, P.F. (1991). Physical models of air flows: a new design tool. In *Atrium Buildings: Architecture and Engineering*, ed. F. Mills, pp. 13–22. Welwyn, UK: Construction Industry Conference Centre Publications.

Bakke, P. & Leach, S.J. (1962). Principles of formation and dispersion of methane roof layers and some remedial measures. *The Mining Engineer*, **22**, 645–69.

Bakke, P., Leach, S.J. & Slack, A. (1964). Some theoretical and experimental observations on the recirculation of mine ventilation. *Colliery Eng.*, **41**, 471–7.

Bedard, A.J., Hooke, W.H. & Beran, D.W. (1977). The Dulles Airport pressure jump detector array for gust front detection. *Bull. Am. Meteorol. Soc.*, **58**, 920–6.

Blackmore, D.R., Herman, M.N. & Woodward, J.L. (1982). Heavy gas dispersion models. *J. Haz. Mat.*, **6**, 107–28.

Britter, R.E. & McQuaid, J. (1989). Workbook on the dispersion of dense gases. *Health & Safety Executive Contract Research Report No. 17/1988*, 119 pp. Sheffield: H.S.E.

Brown, W.G. (1962). Natural convection through rectangular openings in partitions. *Int. J. Heat Mass Transf.*, **5**, 859–68.

Ermak, D.L., Chan, S.T., Morgan, L.K., & Morris, L.K. (1982). A comparison of dense gas dispersion model simulations made with Burro series LNG test spill results. *J. Haz. Mat.*, **6**, 128–60.

Fay, J. A. & Ranck, D.A. (1983). Comparison of experiments on dense gas cloud dispersion. *Atmos. Environ.*, **17**, 239–48.

Fletcher, B., McQuaid, J., & Mercer, A. (1973). Some ventilation problems in the underground environment. *Tunnels and Tunnelling*, **5**, 144–50.

Fujita, T. T. (1981). Tornadoes and downbursts in the context of generalised planetary scales. *J. Atmos. Sci.*, **38**, 1512–34.

Fujita, T. T. (1985). *The Downburst*. Chicago: University of Chicago Press. 122 pp.

Fujita, T.T. & Caracena, F. (1977). An analysis of three weather-related aircraft accidents. *Bull. Am. Meteorol. Soc.*, **58**, 1164–81.

Georgeson, E. M. H. (1942). The free streaming of gases in sloping channels. *Proc. R. Soc. London, Ser. A*, **180**, 484–93.

Leach, S. J. & Thompson, H. (1968). Observations on a methane roof layer at Cambrian Colliery. *Mining Mineral. Eng.*, **August**, 35–7.

McQuaid, J. (1985). *Heavy Gas Dispersion Trails at Thorney Island*. Amsterdam: Elsevier Science Publications.

Mercer, A. (1981). An experimental investigation into the behaviour of a methane accumulation in an internal heading following auxiliary fan start-up. *19th International Conference Research Institute in Safety in Mines, Katowice, Oct. 1981*. pp. 1–8.

Picknett, R. G. (1981). Dispersion of dense gas puffs in the atmosphere at ground level. *Atmos. Environ.*, **15**, 509–25.

Puttock, J. S., Blackmore, D. R. & Colenbrander, G. W. (1984). Field experiments on dense gas dispersion. *J. Hazardous Materials*, **6**, 13–41.

Shaw, B. H. & Whyte, W. (1974). Air movement through doorways. *J. Inst. Heating & Ventilating Eng.*, **42**, 210–18.

Simpson, J. E. & Britter, R. E. (1980). A laboratory model of an atmospheric mesofront. *Q. J. R. Meteorol. Soc.*, **106**, 485–500.

van Ulden, A. P. (1974). On the spreading of a heavy gas released near the ground. *Proceedings of the 1st International Symposium on Loss Prevention and Safety Promotion in the Process Industry, Holland*, pp. 221–5. The Hague: Elsevier.

Wakimoto, R. M. (1982). The life cycle of thunderstorm gust fronts as viewed with doppler radar and rawinsonde data. *Mon. Wea. Rev.*, **110**, 1060–82.

Wheatley, C. J. & Webber, D. M. (1984). *Aspects of the Dispersion of Denser than Air Vapours Related to Gas Cloud Explosions*. Final report of contract between European Energy Community and the UK Atomic Energy Authority.

7   Gravity currents in rivers, lakes and the ocean

Sharp boundaries form in the ocean between adjacent water masses of different density associated with differences in temperature, salinity or amount of suspension and appear at the surface as frontal lines.

Pioneering descriptive and dynamic studies of fronts were made in the seas around Japan, where fronts were classified according to the area where they formed (Uda, 1938). Much descriptive material was gathered from Japanese fishermen, who often noticed a visible line of demarcation at the boundary between water masses. The sea can become quite violent in frontal zones, especially if the current shear across the front is large. Steep pyramidal standing waves may form when wave trains cross into a region of impeding current, or where a strong current flows in a direction opposite to a strong wind.

Flotsam often accumulates along the convergence line, and may include detritus such as dust, foam, timber and examples from the whole food chain up through plankton, mollusca, insects, fish, birds, whales and finally humans (in the form of fishermen and also corpses!)

7.1 Ocean-scale fronts

Many oceanic fronts are formed at the boundaries of two water masses having different origins, and are of sufficiently large scale to be influenced by the earth's rotation.

A large-scale front is formed at the boundary of the northwestern spreading of the gravity current of warm blue Sargasso Sea water flowing over the underlying cold green waters of the North Atlantic. The Gulf Stream is a warm surface current flowing parallel to this boundary, partly due to rotational forces and partly caused by the westerlies and the trade winds. The map in figure 7.1 shows a record of the position of the thermal front of the Gulf Stream during a nine-month period between February and November 1978.

7.1.1 Shallow-sea fronts

Small-scale counterparts of oceanic fronts are found in coastal waters and may result from tidal mixing or from the interaction between fresh and salt water. Figure 7.2 shows a satellite photograph of Britain in which differences in sea surface temperature indicate two such fronts.

The structure of a similar front investigated by Pingree (1974) near the Channel Islands is shown schematically in figure 7.3. This shows two water types A and B separated by a frontal region. Near the front there are convergent motions which sweep all floating material to the edge of the front. Confirmation

Figure 7.1 Temperature discontinuities in the NW Atlantic. The lines show the variability of the thermal fronts of the Gulf Stream through a nine-month period. (After G.A. Maul.)

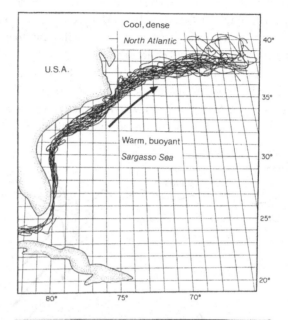

Figure 7.2 Satellite photograph showing fronts in the sea near Britain. (a) Islay–Malin head front, (b) Celtic Sea front. (Courtesy of Dundee University Electronics Laboratory.)

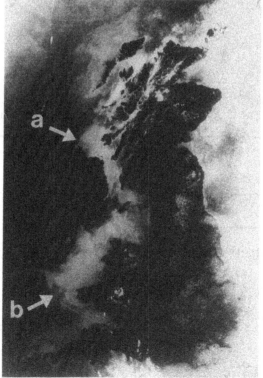

of sinking motions was made by marking the sea surface along one side of the front with a fluorescent dye. This line was seen to move towards the front, deform as shown and disappear beneath the surface.

The flow observed appears to be that of a surface gravity current, as described in section 12.6. The less dense fluid, B, advances relative to the water mass A, forcing surface water beneath it. As the flow from both sides is towards the front, any objects close to the surface will move towards this line, and if sufficiently buoyant will collect there.

These shallow-sea fronts or 'shelfbreak' fronts are commonly formed at boundaries between shallow nearshore waters, which are mixed by winds or tides, and the deeper offshore waters, which remain stratified as shown in figure 7.4. These shelf–sea fronts form in shallow water where tides generate

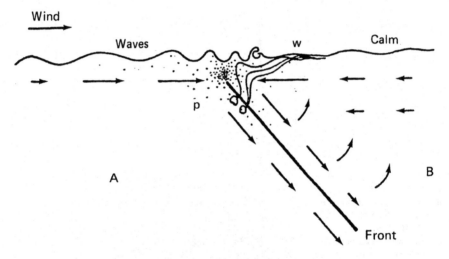

Figure 7.3 A schematic picture of the cross-section of a front separating two water masses A and B. The dots show water-borne material which collects in area w. (After Pingree, 1974.)

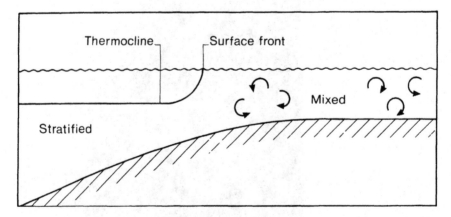

Figure 7.4 The formation of a 'shelfbreak front'. The bottom-generated turbulence prevents the formation of the seasonal thermocline.

enough turbulence at the sea-bottom to prevent the formation of a sharp thermocline.

It has been shown that an important parameter governing this kind of front formation is u^3/h, where u is the strength of the tidal stream and h the depth of the water (Simpson & Bowers, 1981). The rate of work done by the turbulent kinetic energy is proportional to u^3, and the rate of change of potential energy it produces in mixing stratified water is proportional to the depth h. So in areas mixed by strong bottom-generated turbulence, as on the right in figure 7.4, u^3/h is large. In the area on the left, u^3/h is small, and the area remains stratified. Critical values of this parameter have been shown to appear in the frontal areas which appear in the satellite photograph in figure 7.2, namely in the Celtic Sea and near the Scilly Islands.

7.2 Fronts in estuaries

An estuary, the zone of transition between a river and the sea, can have many possible flow regimes. Large tidal stream velocities tend to produce estuaries vertically mixed through the action of bottom-generated turbulence. A high discharge rate of fresh water can lead to significant stratification, reducing the amount of mixing and resulting in a two-layer flow. Frequent new observations of small-scale fronts in estuaries are being made and their biological importance is now being recognised. A comprehensive review by O'Donnell (1993) describes field observations and laboratory experiments, with theoretical models.

One possible configuration in a river with high discharge rate but low tidal currents is illustrated in figure 7.5. Fresh water flows down the estuary over a 'salt wedge' formed by the opposing dense salty current. Both the front of the fresh-water river plume and the salt wedge on the river bed can be maintained throughout the tidal cycle.

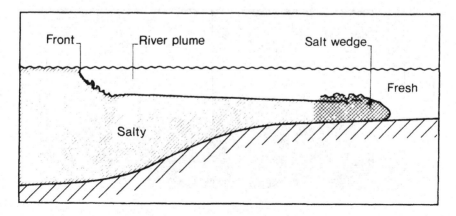

Figure 7.5 River plume and salt-wedge formation in an estuary.

Cross-sections of salt wedges have been measured on some river beds, such as those in the estuary of the Fisher River, which opens into the Strait of Georgia, USA (Geyer, 1983). Figure 7.6 shows how, using echo-sounding, the arrival of a salt wedge was recorded at a station near the coast. As the saline front moved past the station, it displayed a typical gravity current head.

The buoyant plume and dense wedge have been linked together in measurements made in the estuary of the small river Seiont at Caernarfon in North Wales by Simpson & Nunes (1981). The fresh-water discharge is forced back into the estuary during the flood tide, giving the appearance shown in figure 7.7. The tidal inflow comes from the right and passes beneath the river plume forming a front which is clearly marked by floating debris. A characteristic V shape appears in which two frontal arms meet at a point where rapid sinking motions occur.

A two-layer structure extends upstream of the front, along the axis of the river, with about 30 cm of fresh water overlying the intruding salt wedge. During the tidal cycle, after the time of maximum flood, the front starts to move downstream, maintaining the V configuration until the river widens and the front becomes convex. The front may be disrupted, but has a strong tendency to return to the same organised pattern.

Well-documented studies of fresh-water plume fronts have been made by Garvine & Monk (1974) for the Connecticut River, the largest in New England. This river has a high discharge of melt waters from mountains in the north, and in the spring approaches the salt-wedge regime. Figure 7.8 shows an aerial view of the frontal boundary of the Connecticut river plume in April. The river is on the right of the foam line, which sharply divides the yellowish-brown river water from the blue-green waters of Long Island Sound.

Measurements of the vertical density profile along a cross-section of the front are given in figure 7.9. This shows a gravity current of fresh water, which is about half a metre in depth.

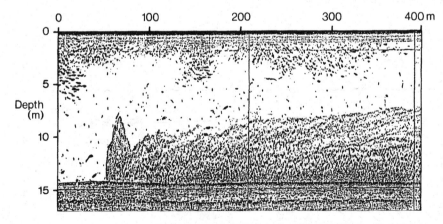

Figure 7.6 Salt wedge in the Fisher River, Georgia, USA detected by echo sounder. (Courtesy of David Farmer.)

Figure 7.7 Tidal intrusion front in the Seiont River, Wales, outlined by floating debris. Saline strait water is moving from right to left beneath the fresh water of the estuary. (Courtesy of R. Nunes.)

Figure 7.8 Aerial view of the frontal boundary of the Connecticut River plume, 26 April 1972. River water is to the right. (Courtesy of R.W. Garvine.)

7.3 Fronts in fjords and lakes

Fjords, or sea-lochs, are deep inlets from the ocean influenced by the tides, the winds and fresh-water input from rivers. Many fjords have a very deep main basin, down to 1700 m, but most have a shallow sill at their mouth, of between 1 and 100 m high.

Several types of gravity current front have been observed in fjords (McClimans, 1978); figure 7.10 shows schematically some of the main processes which form them. Fjords with high river run-off have well-mixed layers of brackish water which can be as deep as 10 m. When the tide is high enough, a dense gravity current may flow over the sill into this brackish water and run down the slope into the fjord. After it has descended some distance it may have the same density as its surroundings; it then starts to move horizontally, forming an interflow or intrusion as shown in the diagram.

The formation of a deep unmixed layer is of practical importance in

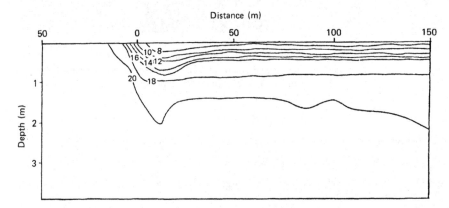

Figure 7.9 Density profile (in parts per thousand) normal to the colour front in the Connecticut River, 14 May 1973.

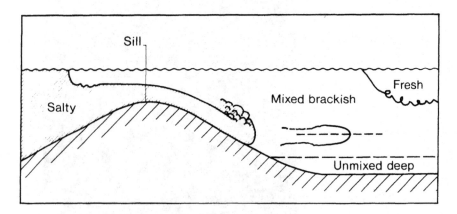

Figure 7.10 Some of the main processes which form gravity current fronts in a fjord.

connection with pollution, and deep water exchange determines the amount of available oxygen, which has important biological significance. The renewal of deep water due to the occasional flow of relatively dense water over the sill may occur only at rare intervals. For example, measurements in Loch Etive and in Loch Eil in Scotland by Edwards *et al.* (1980) have shown a replacement frequency of only about once per year, and that this is due to a gravity current process.

River fronts where extensive foam fronts are observed are permanent features of the surface waters of some fjords.

7.3.1 Lakes and reservoirs

A diagram of the flow of a river into a lake, or reservoir, is shown in figure 7.11. In this case the river carries suspended material and is therefore denser than the water in the lake; it may also be denser due to its low temperature. At the entrance to the lake, the dense river water descends along a clearly marked line called the 'plunge line'. This plunge line is the sign of a stationary front on the surface, between the two water masses. As in surface fronts described above, this front may be detectable from colour changes in the water and by floating material which does not descend in the downward flow at the front. As the river inflow descends the slope in the form of a gravity current, mixing takes place at the head and eventually the fluid may move horizontally as an intrusion or inter-flow.

7.4 Bores in the environment

Chapter 1 introduced bores and outlined some of the features that a bore has in common with the front of a gravity current. They both mark the leading edge of a continued process of mass transfer; this is not a typical feature of the behaviour of waves, the main effect of which is a transfer of energy.

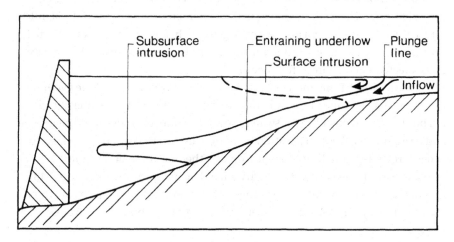

Figure 7.11 The flow of dense river water into a reservoir or lake.

Chapter 13 will examine the dynamics of bores and show how, at a comparatively weak bore, the loss of energy may be effected by a wave train, but that in a stronger bore there must be a turbulent mixing zone at the front. The latter process can be seen at the leading edge of a gravity current (except in very viscous flows), the leading edge being almost identical to that of an internal bore at an interface between two fluids.

7.4.1 Tidal bores in rivers

We showed a picture in Chapter 1 of the bore moving up the river Severn (figure 1.4). This is the best known bore in Britain, and is usually well developed at the spring tides at the spring and autumn equinoxes, when it can be as much as one metre high and may move up the river past Gloucester at speeds between 5 and 7 m s^{-1} (Rowbotham, 1964).

Whether a bore occurs in a river depends on the local conditions at the mouth; two of the most important requirements being a very high tide and a funnel-shaped estuary. Other important factors are the rate of flow of the river and the shape of the river bed (Tricker, 1964).

In Britain there are several other rivers which sometimes produce bores. After the Severn, the Trent usually produces the next most impressive bores. Here the tidal hydraulic jump is named the 'aegre' and the waves following it are called 'whelps' (Barnes, 1952). Clearly detectable tidal bores have been seen in several other rivers, although they may sometimes be only a few inches high.

Over 50 rivers throughout the world are known to have well-formed tidal bores (Bartsch-Winkler & Lynch, 1988). The best developed is probably the one in the Araguari River, on the coast of Brazil, which has been investigated by a team of scientists aboard the Cousteau Society vessel *Calypso*. The photograph in figure 7.12 shows this to be an undular bore, with over 20 waves clearly visible behind the front of the bore.

Some of the South American bores have only recently been explored, but in China one of the largest bores in the world has historical records going back over 2000 years. The Qiantang River in China, which enters the sea only 50 miles south of the Yangtse, has a bore with a legendary history going back to the fifth century BC. Already in the second century BC the Chinese understood its nature and expected a good bore at full moon.

As the city of Hangzhou grew up between the West Lake and the Qiantang River, its safety became more and more precarious and in 910 AD Governor Qian Liu built a dyke to meet the tide-water. He made bamboo bands, piled huge stones and drove in large trees. In addition he stationed hundreds of crossbowmen to shoot arrows, nominally 'to stay the forces of the tide', but perhaps more effectively as a public relations activity. An old print reproduced in figure 7.13 gives an impression of the scene as they began to build the wall, parts of which still stand today. The bore can be seen advancing as a turbulent line from the left.

Throughout the following thousand years the bore continued to trouble

Figure 7.12 Tidal bore at the mouth of the Araguari River, in Brazil. The bore is undular, with over 20 waves visible. (Courtesy of D.K. Lynch.)

Figure 7.13 Protection against the bore in the Qiantang River: an historic scene.

the people of Hangzhou. It seems to have been somewhat variable in strength, since Marco Polo did not remark on it during his visit.

In the nineteenth century careful records were made by a British naval officer (Moore, 1888), who recorded bores of up to two metres in height. He noted how navigation was effected by the use of 'bore shelters'. These consisted of shelves half-way up the river wall, well above the water level, on which vessels were perched until the bore arrived and floated them off.

Although work had been done on maintaining the walls and digging out the channels, severe bores still sometimes occurred less than 20 years ago, in the Qiantang River. An example is shown of one of these in figure 7.14. In these views a special feature is the dangerous intensification which can occur as the result of reflection of a bore which strikes the wall at an angle. The turbulent bore

a

b

Figure 7.14 Intensification of a bore by reflection. In (a) the Qiantang bore in China has been reflected from the wall seen to the left. In (b) the original and reflected flow meet at the bend in the wall. (Courtesy of Zhejiang Provincial Estuarine Research Institute.)

approaches from the top right-hand side and is reflected from the wall on the left. After reflection the height of the bore is doubled and it strikes the obstacles in the foreground with great force. Work carried out by the Zhejiang Provincial Estuarine Research Institute has reduced the effects of the bore; this has been done by deepening the river channels.

The highest tides in the world occur in the Bay of Fundy in Canada and, as might be expected, some well-known river bores are associated with them (Dalton, 1951). One of these rivers, the Petitcodiac, has a very impressive bore which may be seen to best advantage from the riverside park provided for this purpose in the city of Moncton, New Brunswick.

7.5 Internal bores

There is good evidence for the existence of internal bores in rivers, lakes, fjords and the oceans. Echo-sounders can display the bore profiles and detailed measurements have been made using sensors of temperature, salinity and velocity.

These bores are usually generated in a comparatively thin layer of water which lies above a deeper denser layer, so the front of the bore would be expected to appear as a wave of depression. Several different mechanisms have been established for the formation of internal bores in these commonly existing stratified layers.

7.5.1 Surges in stratified lakes

Internal bores have been observed in Loch Ness, which is 35 km long, about 2 km wide and 200 m deep (Thorpe *et al.*, 1972). The bores propagate in undular form at speeds up to 0.2 m s^{-1} towards the north west. They are often 10 m deep, with a series of waves to the same depth, about 10 km long.

The form of the bore is best shown in temperature measurements, as in figure 7.15. This record was made from a moored boat from which a probe was suspended at a depth of 40 m near the centre of the loch. The increase in temperature as the front of the bore passed shows that the isotherms were lowered and that the bore was a wave of depression. The change in level of the isotherms was about 12 m and the mean speed of propagation of the bore was found to be 0.37 m s^{-1}. It is believed that this bore is forced by wind-stress causing flow of near-surface water to one end of the loch.

Figure 7.15 Temperature at 40 m depth in Loch Ness during the passage of internal bore. (After Thorpe *et al.*, 1972.)

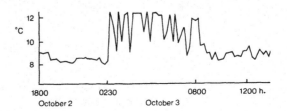

Even larger undular bores have been measured in Seneca Lake, New York by Hunkins & Fliegel (1973), where the isotherms are often 20 m deeper behind the front of the bore, which travels at 0.35 to 0.40 m s^{-1}.

7.5.2 Internal tidal bores in the ocean

Internal bores have been investigated off the coast of California by Cairns (1967). During the summer months, when a strong seasonal thermocline exists, the thermal structure approximates a two-layer system. It was found that as an internal tide moved inland its wave profile became increasingly asymmetric as the wave entered shoaling coasting waters. This asymmetry became more pronounced with increasing wave height and resulted in the formation of an internal tidal bore.

In other measurements made at La Jolla, California (Winant, 1974) similar surges of cold water were recorded. Since the surge was evident even on the bottom it was believed to be due to the run-up of broken internal waves.

7.5.3 Flows over sills or through contractions

The presence of a sill near the mouth of a river or fjord may be responsible for the formation of an internal bore at certain times during the tidal cycle. In many cases a response to tidal flow over a submarine sill is an internal hydraulic jump downstream of the sill (Gargett, 1980). Such a flow occurs in Knight Inlet, British Columbia, on the turn to flood tide when an undular bore emerges from the region of the sill. Figure 7.16 shows a record from a 200 kHz hull-mounted echo-sounder, taken as the ship traversed the bore. The leading edge of the undular bore is to the left of the picture and typical wavelengths are of the order of 100–200 m near the leading edge. At this internal jump the height ratio, or strength, is about 2.

Undular bores form at certain tidal stages in the two-layer flow through the Straits of Gibraltar (Farmer & Armi, 1986). Since in the Mediterranean, the effect of evaporation exceeds that of river discharge, the water is more saline than that of the Atlantic. As the tide rises, a layer of less-saline water flowing above a sill in the Strait enters the Mediterranean as an internal bore. The satellite photograph

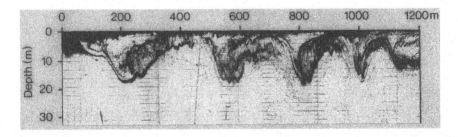

Figure 7.16 Echo-sounder record of an undular internal bore in Knight Inlet, British Columbia. (Courtesy of David Farmer.)

in figure 7.17 shows interfacial features associated with internal flow glinting in the sun and marks clearly the front of the internal undular bore spreading to the right into the Mediterranean.

7.5.4 Generation of internal bores by gravity currents
A number of observations have been described which are consistent with the generation by gravity currents of internal bores near water surfaces. Unfortunately, many of these are not well documented.

Developing sets of lines at the surface suggesting an undular bore have been recorded on several occasions. Some of these have appeared as foamlines on the surface and others, from airborne thermal imagery, have shown periodic lines of variation of surface temperature.

Figures 7.7 and 7.8 have shown how the convergence zone at the front of a gravity current of less-dense fluid can cause a single sharp line of floating foam and debris. Laboratory experiments can reproduce the early stages in the formation of a bore at the front of a gravity current. As the front of the bore moves along the surface of the water ahead of the leading edge of the gravity current it will develop its own separate foam line. This process will be repeated until a series of parallel foam lines is formed. Figure 7.18 is a photograph of a regular series of foam lines in Trondheim Fjord, Norway (McClimans, 1978), which is thought to have this explanation.

A similar pattern, suggesting an internal bore, has also been found in fresh-water river discharges into the sea. Cold fresh water from the Quinault River discharging into the NE Pacific shows a series of lines which have been shown to be successive boundaries of warm coastal–ocean water (Gross, 1972).

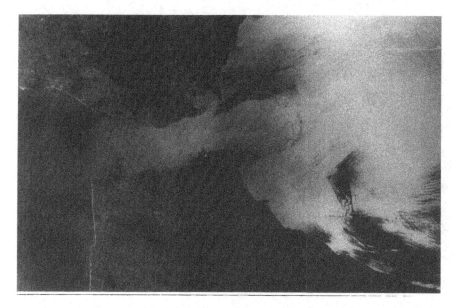

Figure 7.17 An undular bore formed by tidal flow through the Straits of Gibraltar. (Satellite photograph, courtesy of NASA).

7.6 Turbidity currents on the ocean bed

The deep ocean floor is sometimes violently disturbed by turbidity currents moving down from the continental shelf. The attention of oceanographers was first drawn to this fact by breaks in submarine telephone cables, but the suggestion that turbidity currents were responsible for these breaks and other phenomena such as submarine canyons formerly created much controversy.

Some enormous submarine 'landslides' are known to have set up turbidity currents at the edge of the continental shelf. The volume of sediment deposited by the well-studied turbidity current off the Grand Banks near Newfoundland in 1929 has been estimated to be as much as 100 km^3. This is a thousand times the volume of the greatest landslide recorded on land. The events started with a severe earthquake which immediately broke a large number of submarine cables near its epicentre on the continental shelf just south of Newfoundland. But none of the numerous cables crossing the continental shelf in the area further south were affected. Further downstream from the epicentre, however, another cable broke an hour later. Then four other cables broke in succession, each in two or three places, the last break occurring 13 hours later and 300 miles distant. The details are shown in figure 7.19.

Submarine canyons are found in many sections of the continental slope. They are similar to river gorges on earth, with equally steep sides and depths. Turbidity currents have been observed actually flowing down some of these

Figure 7.18 A regular series of foam lines seen in Trondheim Fjord. (Courtesy of the Norwegian Hydrotechnical Laboratory.)

canyons. A good example is the event in 1935 at the mouth of the Rio Magdalena in Colombia, when 450 m of a breakwater suddenly disappeared into the sea, and the same night a submarine cable was broken at a point 24 km out to sea and 1.5 km deep in one of the canyons which extend out to sea from near the river mouth. During repairs the cable was found to have grass wrapped round it of the type growing near the breakwater, suggesting that the slump had developed into a turbidity current capable of producing this long-distance transport. In this district, during the first 25 years since one of the cables was laid it has been broken by turbidity currents on 17 occasions.

Abyssal plains occur in many oceans; for example, they cover 10% of the floor of the North Atlantic, deviating from smoothness by only a few metres. Echo-soundings show, however, that the smooth plains are themselves underlain by a rugged base. Analysis of cores from these plains shows that they consist of sand and silt layers which appear to have come from the continental shelf, and that transport by turbidity currents is the most likely explanation.

Figure 7.19 Submarine cables broken by a turbidity current from the Grand Banks sediment slide, 1929 (marked by the dotted area).

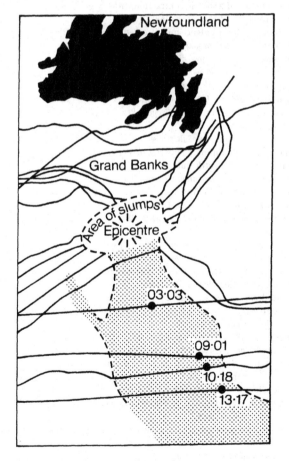

Bibliography

Barnes, F.A. (1952). The Trent Aegre. *Survey, University of Nottingham*, **3**, 1–16.

Bartsch-Winkler, S & Lynch, D.K. (1988). *Catalog of Worldwide Tidal Bore Occurrences and Characteristics. US Geological Survey Circular 1022.* Denver: Federal Center, 17 pp.

Cairns, J.L. (1967). Asymmetry of internal tidal waves in shallow coastal waters. *J. Geophys. Res.*, **72**, 3563–65.

Dalton, F.K. (1951). Fundy's prodigious tides and Petitcodiac's tidal bore. *J.R. Astron. Soc. Can.*, **45**, 225–30.

Edwards, A., Edelsten, D.J., Saunder, M.A. & Stanley, S.O. (1980). Renewal and entrainment in Loch Eil, a periodically ventilated Scottish fjord. In *Fjord Oceanography*, ed. H.J. Freeland *et al.*, pp. 523–34. New York: Plenum Press.

Farmer, D.M. & Armi, L. (1986). Maximal two-layer flow over a sill and through the combination of a sill and contraction with barotropic flow. *J. Fluid Mech.*, **164**, 53–76.

Gargett, A. (1980). Turbulence measurements through a train of breaking internal waves in Knight Inlet, BC. In *Proceedings of the NATO Conference on Fjord Oceanography, Victoria, BC*, pp. 277–81. New York: Plenum Press.

Garvine, R.W. & Monk, J.D. (1974). Frontal structure of a river plume. *J. Geophys. Res.*, **79**, 2251–9.

Geyer, R. (1983). *Fraser River Salt Wedge Investigation*. Preliminary Report, University of Washington, Seattle, July 13, 1983.

Gross, M.G. (1972). *Oceanography: A view of the Earth*. New Jersey: Prentice Hall.

Hunkins, K. & Fliegel, M. (1973). Internal undular surges in Seneca Lake. *J. Geophys. Res.*, **78**, 539–48.

McClimans, T.A. (1978). Fronts in fjords. *Geophys. Astrophys. Fluid Dynam.*, **11**, 23–34.

Moore, W.U. (1888). The Bore of the Tsien-Tang Kiang (Hang-Chau Bay). *J. China Branch, Royal Asiatic Society, Shanghai*, **23**, 185–247.

O'Donnell, J. (1993). Surface fronts in estuaries: a review. *Estuaries*, **16**, 12–39.

Pingree, R.D. (1974). Turbulent convergent tidal fronts. *J. Mar. Biol. Assoc. UK.*, **54**, 469–79.

Rowbotham, F.W. (1964). *The Severn Bore*. London: David & Charles. 100 pp

Simpson, J.H. & Bowers, D. (1981). Modes of stratification and frontal movement in shelf seas. *Deep-Sea Res.*, **28A**, 727–38.

Simpson, J.H. & Nunes, R.A. (1981). The tidal instrusion front: an estuarine convergence zone. *Estuarine, Coastal Shelf Sci.*, **13**, 257–66.

Thorpe, S.A., Hall, A., Crofts, I. (1972). The tidal surge in Loch Ness. *Nature*, **237**, 96–8.

Tricker, R.A.R. (1964). *Bores, breakers and waves.* London: Miller & Bonn.

Uda, M. (1938). Researches on 'Siome' or current rip in the seas and oceans. *Geophys. Mag.*, **11**, 307–72.

Winant, C.D. (1974). Internal surges in coastal water. *J. Geophys. Res.*, **79**, 4523–6.

8   Industrial problems with gravity currents: oceanography

8.1 Power station effluents

Fossil fuel and nuclear power stations raise steam to drive turbines. Their efficiency is less than about 40% and the balance of the heat produced is rejected to the condenser cooling water; if it is impracticable to make use of this heat, it must be dissipated in the environment.

At coastal stations the obvious method of cooling the power stations is by direct, or once-through, cooling. Water is pumped from the sea, through the station condensers where it takes up latent heat from the condensing turbine steam, and is returned to the sea from the cooling water outlet.

The cooling-water requirement per GW of generating capacity is about $30 \ \text{m}^3 \text{s}^{-1}$ or $50 \ \text{m}^3 \text{s}^{-1}$ for a nuclear plant (Macqueen, 1978). The fresh-water flow in the River Thames at Teddington Lock averages $75 \ \text{m}^3 \text{s}^{-1}$ so that the cooling water from a 2 GW power station can be compared with a fairly large natural river.

This cooling water from the sea is heated up about 10 °C and discharged back again. It is important to know where the discharged water goes and how quickly it cools down, both for ecological reasons (the discharge might affect marine life) and for station efficiency (recirculation, i.e. the reuse of already warmed water, lowers efficiency).

When the cooling water leaves the outlet, it tends to rise to the surface and spread out horizontally. Very close to the outlet, in the so-called 'near-field', the discharge is dominated by the momentum of the flow.

A few hundred metres further away, in the 'mid-field', the discharge behaves much like a gravity current. That is to say, sharp boundaries are maintained between the outflow and denser surrounding water, and these are maintained even when the flow is disturbed by variations of depth or turbulence in the surrounding water.

In the 'far-field', turbulence processes dominate and no sharp horizontal boundary can be seen between the flow and the ambient water.

Figure 8.1 sums up what is thought to happen to a radially symmetrical rising jet in deep still water, showing the near- and mid-field. The rising jet entrains surrounding water and spreads horizontally. If this flow is still supercritical, then an internal hydraulic jump is expected, leading to the gravity current head shown at the front of the flow. The typical value of diluted gravity, g', at the outflow is $0.025 \ \text{m}^3 \text{s}^{-1}$ for a 10 °C rise in temperature, leading, for example, to a spreading velocity of 6 cm s⁻¹ (or 200 m h⁻¹) for a 1 m deep current, of the order of the spreading observed.

British coastal waters are tidal and the spread of the cooling-water plume is very sensitive to the ambient current. Extensive research on cooling water in tidal flows has been carried out at Sizewell power station on the east coast of England (Ewing, 1982).

Figure 8.2 gives an interpretation of such a flow during the flood tide, with vertical scales exaggerated. There is considerable mixing in A, the initial stage during the rise from the outfall ports. In B, the stable gravity current stage, there is less vertical mixing. In the far-field, C, the buoyancy forces no longer dominate and heat is transported vertically by turbulent processes. In figure 8.2, as in most models of this process, the plume is shown to thicken with increasing distance away from the outfall due to a combination of interfacial shear entrainment and mixing caused by the turbulent surroundings. Some recent models have a different viewpoint of turbulent mixing (Rodgers, pers. commun.), predicting that turbulence in the ambient water could have the effect, not of diffusing the warm water, but of restraining it in a well-defined layer, while the sub-plume layer gradually warms up fairly uniformly through its layer. Experiments described in Chapter 14 give some idea of how these two processes can come about.

8.2 Oil slicks

The spreading of oil over the surface of the sea is an environmental process in which gravity currents play an important part. As oil production and oil transport over the sea continue to increase, the public have become familiar with the frequent damage caused by oil spillage. A few spills of the order of 100 000 tons of

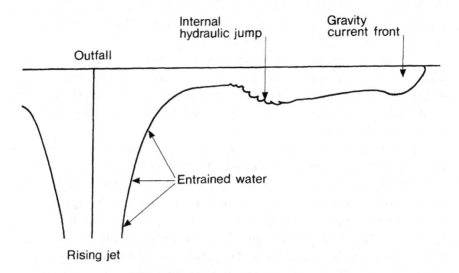

Figure 8.1 A rising jet over deep still water.

crude oil have focused world-wide attention on the problem and there has been much anxiety over the problems of where the oil goes and of the area covered.

Firstly, winds and ocean currents move the oil mass as a whole. Secondly, since the oil has a density 10 to 20% less than water, gravity-controlled spreading is very significant in the behaviour of the oil.

The initial growth of the slick is similar to the early stages of any gravity current immediately after release, and the behaviour which follows has been investigated both theoretically and in the laboratory. Figure 8.3, taken during a

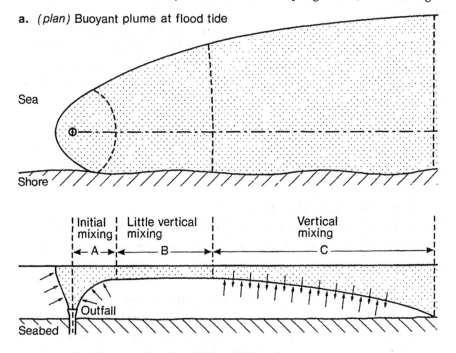

a. *(plan)* Buoyant plume at flood tide

Sea

Shore

Initial mixing · Little vertical mixing · Vertical mixing

← A →|← B →|← C →

Outfall

Seabed

b. *(section)* Mixing of plume water parallel to shore

Figure 8.2 Plan (a), and elevation (b) of a cooling-water plume. A, initial stage; B, gravity current stage; C, far-field.

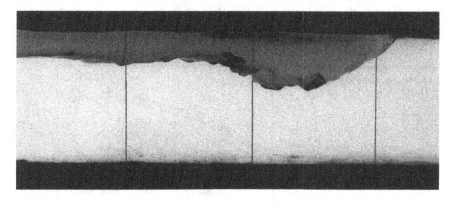

Figure 8.3 A gravity current of an immiscible fluid, advancing on the surface of water. The lines are 10 cm apart.

laboratory experiment, shows the front of a gravity current of immiscible fluid (kerosene) flowing above water. The flow here is in the gravity–inertial regime and shows a turbulent zone behind the deepened head of the current.

The flow which follows the initial collapse can be divided into three regimes, gravity–inertial, gravity–viscous and viscous–surface (Hoult, 1972: Fannelop & Waldman, 1971). Figure 8.4 shows a time-scale for these regimes in the development of a typical oil slick, the depth of which will be of the order of a centimetre.

The equations show that in the gravity–inertial regime, the viscous drag at the oil–water interface is negligible; in the gravity–viscous regime, the thickness is small compared with the previous regime and the drag of the water on the slick becomes dominant; and after a certain time oil spilled on the sea ceases to spread; the process ends in a surface-tension regime.

Time-scales may be:

gravity–intertial   1000 seconds
gravity–viscous   up to a week
viscous–surface   a week to a month

8.2.1 Containment of oil slicks

Floating booms have been used to try to contain oil on a water surface, but their effectiveness is limited by currents, winds and waves. A vertical barrier is towed by ships or fixed in the stream, and the thin layer maintained is removed by skimmers.

An oil boom can fail by the mechanism of oil underflow when oil is swept under the boom by currents. The forms of oil slicks arrested by booms have been investigated experimentally (Wilkinson, 1972) and it has been shown that oil will tend to build up against such a barrier. A slick will propagate upstream as increasing quantities of oil are contained and three possible outcomes are illustrated in figure 8.5.

In case (a) the oil–water interface of the slick is stable and, if the upstream front is far enough away, interfacial disturbances there will die away and a stable wedge can be contained by the barrier.

Figure 8.4 Combined spreading laws. (Courtesy of D.P. Hoult.)

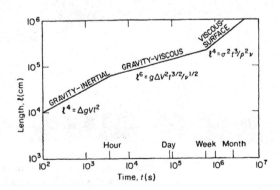

The second type of slick in (b) may occur when the barrier is not deep enough to retain the oil in an otherwise stable slick. The oil will continually drain beneath the barrier, and in time nearly all the oil will be lost.

The third type of slick (c) has no equilibrium thickness and the waves and mixing at the interface make retention of the slick impossible. The dynamic instability which makes containment impossible occurs when the Froude number of the flow upstream of the slick exceeds a critical value and a gravity current head-wave is formed. In flows with Froude number above this value turbulent eddies will form at the oil–water interface and oil will be lost beneath the boom no matter how deeply it is immersed. Figure 8.6 shows how the separated oil droplets are swept past the barrier in the current.

The form of the oil barrier is usually a catenary, as shown in figure 8.7, and the oil pool is found to have a nearly straight leading edge. Along the centre line,

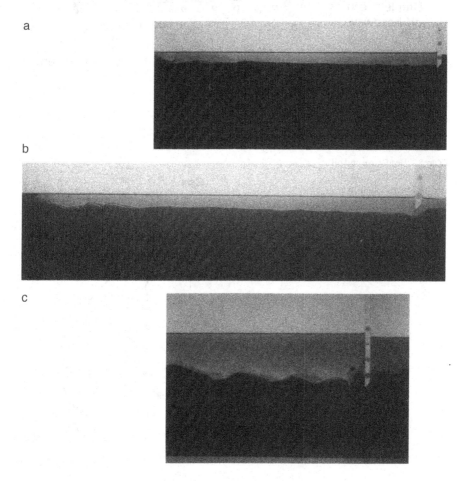

Figure 8.5 Laboratory experiments of oil slicks. (a) Oil slick contained by barrier. (b) Oil escaping beyond barrier of insufficient depth. (c) Oil slick with unstable interface. (Courtesy of D.L. Wilkinson.)

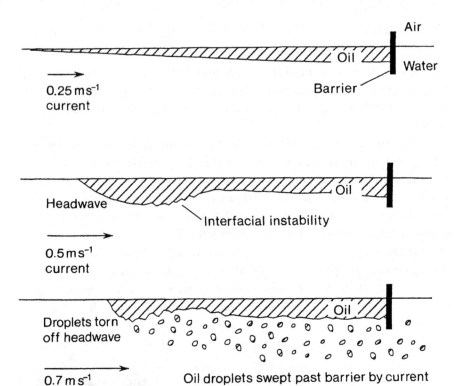

Air

Oil

Water

Barrier

0.25 ms⁻¹
current

Headwave

Interfacial instability

Oil

0.5 ms⁻¹
current

Droplets torn
off headwave

Oil

0.7 m s⁻¹
current

Oil droplets swept past barrier by current

Figure 8.6 Head-wave formation and oil losses due to
entrainment with increasing current speed.

Figure 8.7 Head-wave and oil
barrier geometry, of an oil
slick of length *L*, held at a
barrier by water of flow *U*.

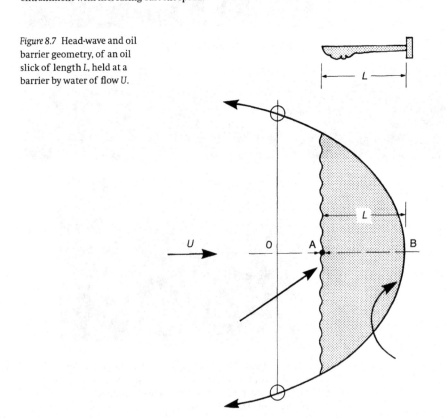

AB, the flow is nearly two-dimensional and the results of laboratory experiments in channels can be applied here, as very few three-dimensional detailed measurements are available.

It is primarily the head-wave and the near-boom region which are responsible for losses and limit the performance of the boom – a detailed discussion of the problems is given by Fannelop (1983).

Bibliography

Ewing, D.J.F. (1982). The spreading-out of cooling-water discharged from direct-cooled power stations. *Central Research Laboratories Report*. Leatherhead: CEGB.

Fannelop, T.K. (1983). Loss rates and operational limits for booms used as oil barriers. *Appl. Ocean Res.*, **5**, 80–92.

Fannelop, T.K. & Waldman, G.D. (1971). The dynamics of oil slicks. *AIAA Journal*, **10**, 506–10.

Hoult, D.P. (1972). Oil spreading on the sea. *Ann. Rev. Fluid Mech.*, **4**, 341–68.

Macqueen, J.F. (1978). Background water temperatures and power station discharges. *Adv. Water Res.*, **1**, 195–203.

Wilkinson, D.L. (1972). Dynamics of contained oil slicks. *J. Hydr. ASCE*, **98HY6**, 1013–30.

9   Avalanches

9.1 Types of avalanche

Many different cases of mass transport occurring under gravity have been
studied in geological science. A well-known example of these is the snow
avalanche, which has already been mentioned in Chapter 1. There are several
other types of avalanche; one of the most significant of these is an avalanche
of debris or mud which can be even more destructive and dangerous.

The whole range of debris flows represents a set of processes lying
between the drier forms of mass transport and liquid stream flow. At the latter
end of this range, mud flows can occur. These have a high content of water and
may be derived from hill slopes which have disintegrated with catastrophic
effect.

Volcanic eruptions are responsible for several types of gravity-controlled
flows including debris and mud flows. Two of these are uniquely volcanic –
pyroclastic flows and streams of basaltic lava. These will be dealt with in detail in
Chapter 10.

9.2 Snow avalanches

People living in alpine regions are exposed in winter and spring to avalanches of
snow, and research on the dynamics and prevention of avalanches is carried out
by most of the mountainous nations. There are many historical records of snow
avalanches, for example of their creation by the passage of Hannibal's men and
animals across the Alps in 218 BC.

An early illustration of an avalanche is shown in Figure 9.1, a wood
engraving by H. Schaufelein in the 'Theurdanck' of 1517. This pictures an
avalanche as a series of large snowballs; as late as the end of the nineteenth
century people continued to imagine avalanches as giant snowballs which
increased by accretion of the underlying snow as they bounded down the
mountain side.

Since about 1900 organised scientific research on avalanches has developed
in many countries, notably in Switzerland, France, Japan, Canada and the USA,
and its importance is still growing. The areas of practical importance in this
study can be divided into three: the starting zone, the avalanche track and the
run-out area.

9.2.1 Starting zone

First, in the 'starting zone', there is the problem of the release mechanism. Snow is not a simple material and the mechanism of the initial fracture is almost impossibly complex theoretically. In spite of the large body of information on the properties of snow that can be found in a recent review (Salm, 1982), prediction of avalanche releases remains essentially empirical. The snow cover may break away to a depth of the order of one metre to generate an avalanche and this may be caused by changes of the structure of the snow. Other possible causes of avalanche triggering are ski loads, falling cornices and earthquakes. They are also initiated artificially.

9.2.2 Types of snow avalanche

In the second stage, the mature avalanche travels down the slope as a gravity current. Snow avalanches may be crudely classified into two types, flow avalanches and airborne powder snow avalanches. Most avalanches do not belong to either of these limiting cases, but are rather a mixture of the two. The two limiting cases are shown in figures 9.2 and 9.3, respectively. Wet-snow avalanches form a class of flow avalanches and have some interesting features in common with debris flows and mud avalanches, to be dealt with in section 9.4.

Flow avalanches have a typical velocity of up to 60 m s^{-1}, and a flow height of 5 to 10 m. Powder snow avalanches move at velocities over 100 m s^{-1} and can be over 100 metres high. The graph in figure 9.4 shows a set of observations from field work of the velocity of both powder snow avalanches and flow avalanches (Hopfinger, 1983), plotted against the depth of the flow.

A flow avalanche initially tends to slide like a rigid body but rapidly breaks

Figure 9.1 Sixteenth century illustration of a snow avalanche.

up into smaller particles and turns into a granular material flow. The
'fluidisation' mechanism for such a flow depends on the interchange of energy
of rapidly colliding small particles, and will be discussed again in the following
section on dry rock avalanches.

An airborne powder snow avalanche is essentially a turbidity current
flowing down an incline. The behaviour of slope gravity currents and turbidity

Figure 9.2 Flow avalanche.

Figure 9.3 Powder snow
avalanche. (Courtesy of G.
Kappenberger.)

currents in the laboratory will be described in Chapters 12 and 16. Additional work to clarify the entrainment process in powder snow avalanches has been carried out, together with numerical models of the flows (Scheiwiller, 1986).

Mechanisms are illustrated in figure 9.5 by which the suspension of snow particles is built up, to form a cloud 100 m deep. The turbulent flow over the ground may result in entrainment of further fresh snow from beneath and more air will be entrained above the head.

It is generally agreed that powder snow avalanches develop from flow avalanches, but the mechanism by which the transition from one regime to the other takes place is not clearly understood. A trailing cloud often develops

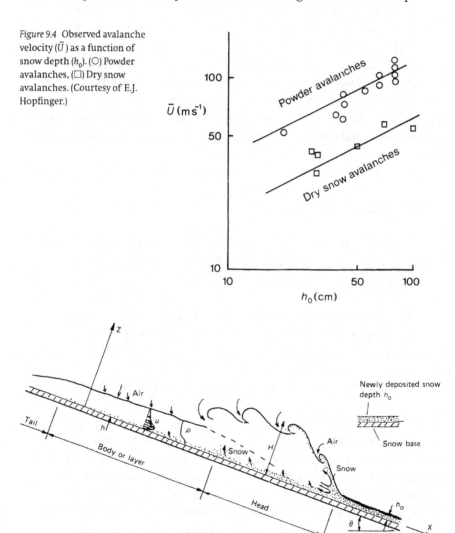

Figure 9.4 Observed avalanche velocity ($\bar{U}$) as a function of snow depth (h_0). (O) Powder avalanches, (□) Dry snow avalanches. (Courtesy of E.J. Hopfinger.)

Figure 9.5 Parts of an avalanche. A large cloud of snow dust lies above the turbulent zone of dense snow. H is the height of the avalanche, h that of the tail and h_0 that of the undisturbed snow. Velocity u and density ρ are shown.

behind a flow avalanche which may increase in volume and actually overtake the dense snow body if the track is long. In these cases a powder snow avalanche may form very rapidly, as the air entrainment grows explosively.

9.2.3 Run-out zone

It is important in practice to know when an avalanche will come to a stop and it has been shown that obstacles can reduce the run-out distance (Salm, 1966; Hopfinger & Tochon-Danguy, 1977). In contrast to flow avalanches, powder snow avalanches behave like Newtonian fluids and flow around an obstacle. Generally, a fast-moving dry snow avalanche is accompanied by a cloud riding above it; when the avalanche hits a barrier the dense portion is stopped and the cloud behaves like a fluid flowing around an obstacle. A low barrier will usually have the effect of accelerating the flow just above it, and damage can be increased. Figure 9.6 shows the results of a related laboratory experiment in which the flow is shown passing a barrier. Although there are strong eddies behind the obstruction the results show an appreciable protected zone.

9.3 Rock avalanches (dry)

Small rockfalls can be derived from soil creep or freeze–thaw, but the larger ones are usually caused by earthquakes. After a short transitional stage the rock fall may develop into an avalanche of rocks of assorted sizes which forms a fluidised flow.

Rockfalls of relatively small volume occur frequently in mountainous regions, and the inhabitants have learned to live beyond their range. Occasionally, however, rockfalls of unusually large volume occur. The material from these travels long distances, perhaps many kilometres beyond what is expected of a mass of solids sliding down an inclined plane with a normal coefficient of friction.

In 1881 the Alpine village of Elm in Switzerland was virtually wiped out by one of these enormous rockfalls, which flowed 2 km down the valley. This disaster made geologists and engineers aware that large masses of rock debris may sometimes behave like a fluid with a low internal resistance. Since 1881 many other instances of fluidised rock avalanches have been found, both contemporary and historic. Such mobile debris streams, sometimes known as 'Sturzstroms' have been extremely costly in human lives and despite many investigations, no universal explanation has been agreed for the extremely long travel distances of rock avalanches.

Figure 9.7 shows the travel distance of a Sturzstrom against the volume of rock deposited (Davies, 1982). This plot is for volumes greater than 10^7 m^3, the range associated with unexpectedly long travel distance. These results strongly suggest that the deposit extent of a rock avalanche depends mainly on its volume, and not the height through which it has fallen. This suggests a fluid-like spreading rather than a sliding mechanism.

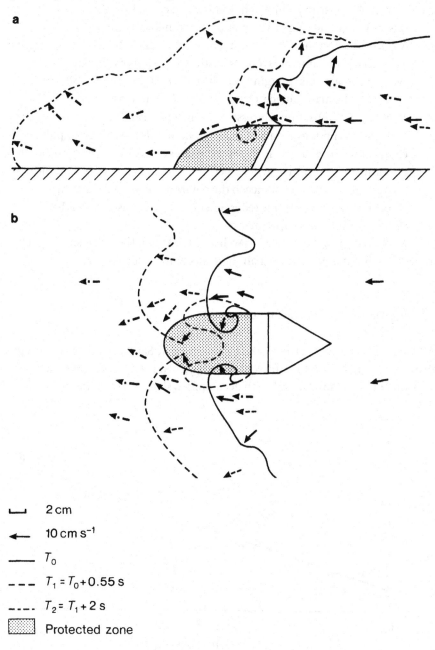

⊔	2 cm
⟵	10 cm s^{-1}
——	T_0
- - -	$T_1 = T_0 + 0.55$ s
-·-·-	$T_2 = T_1 + 2$ s
�	Protected zone

Figure 9.6 The flow of an avalanche past an obstruction. (a) side view, (b) top view. (Courtesy of E.J. Hopfinger.)

Possible explanations in terms of proposed fluidisation mechanisms include:

1.   Air fluidisation. This cannot explain all the observations, because similar flows have been inferred from deposits on the airless surfaces of the Moon and Mars.
2.   Basal melting theory. Much-reduced friction could be accounted for if there was sufficient melting of the rock at the points of contact.
3.   Acoustic fluidisation. Fluidisation by sound energy is believed to be possible through the medium of strong acoustic waves, or 'noise' within the avalanche mass. It is thought that sufficiently large amounts of sound energy can be maintained in large enough flows (Melosh, 1980).
4.   Mechanical fluidisation. The essence of mechanical fluidisation is that a high energy input to a mass of granular material causes high impulsive contact pressures, due to high shearing rates as it moves over the ground beneath. As a result the particles become randomly separated and the mass dilates: the internal resistance to the deformation of the material as a whole is thus reduced. This reduction of internal resistance has been shown in laboratory experiments.

Mechanical fluidisation seems well able to explain the behaviour of rock avalanches, but it is possible that all these mechanisms play a part.

9.4 Debris and mud avalanches (wet)

After a visit to the Alps in the 1780s the naturalist de Saussure wrote about 'the danger of being surprised by torrents which descend with incredible velocity … a kind of liquid mud mixed with decomposed slate and rock fragments; the

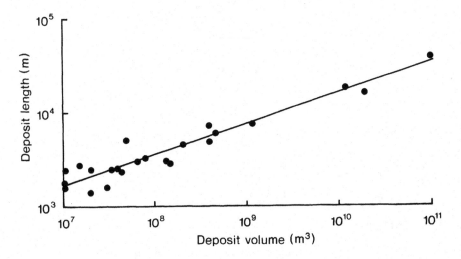

Figure 9.7 Deposit length of rock avalanches plotted against deposit volume. (After Davies, 1982.)

impulsive force of this dense paste is incomprehensible; it incorporates rocks, topples the buildings which happen to be in its way, uproots the tallest trees, and upon bursting forth from ravines ravages the fields, covering the soil with a considerable thickness of silt, gravel and rock fragments.'

Indonesia has suffered from frequent volcanic mudflows, and the Indonesian word for them, *lahars*, is often used to describe them. Many mud avalanches of all sizes, frequently not connected with volcanoes, have been described since de Saussure; one particularly distressing event happened in 1966 in Wales as the result of a comparatively small mud flow. About 140 000 m³ of coal-mining waste, which had become sodden after heavy rain, collapsed abruptly, and a black tongue of coal dust, shale and water flowed down the valley floor into the village of Aberfan. It engulfed 18 houses and the school before it came to rest, killing a total of 144 men, women and children.

The speed at which these flows travel depends on many things, including the steepness of the slope and the mass and viscosity of the slurry. It can be very fast indeed, of the order of 90 km h^{-1}. Values of the kinematic viscosity for mud flows have been estimated as being between 2×10^3 and 6×10^3 cm² s^{-1}, as compared with the value of 10^{-2} for water at 20 °C.

Alluvial fans consist of deposits of coarse and fine debris which are brought down from mountain ranges and spread out on the plain. Only occasional reports have been given of the behaviour of the actual mud flows which are their origin. A graphic description has been given of the appearance of a debris and mud flow as it descended from the White Mountains, on the border of California and Nevada, contributing to an alluvial fan at the base of the mountains. The following events were reconstructed from interviews with two stockmen whose ranches were at the edge of the fan (Beaty, 1963).

1. Two hours after a heavy thunderstorm in the mountains on the afternoon of July 26, 1952, loud rumbling and roaring noises were heard emanating from the lower canyons of the affected drainages.
2. About 30 minutes later, masses of debris were noticed advancing downslope on the upper parts of the fans. The leading edge appeared to be a low wall of boulders and thick mud, without visible water.
3. The debris appeared to be advancing in a series of waves or surges, each wave overtaking and submerging the preceding one.
4. The flows were accompanied by noise likened to 'the sound of thousands of freight cars bumping together simultaneously'.
5. At the lower part of the canyon the material moved 'about as fast as man can dog-trot', perhaps 400 or 500 feet per minute. It was 1 to 2 feet thick.

The photographs in figure 9.8 show smaller-scale debris flows as developed on the cone of sediment rejected from a gravel washing pit. The flow in (a) shows the formation of a levee at the sides of the upper slope and the spreading out in the plain as the gradient decreases. That in (b) shows clearly the overhanging snouts of the flow, which is only a few centimetres thick.

Dam-bursts have often resulted in catastrophic mudflows; one such disaster occurred in the Italian Dolomites in 1985 when 260 people were killed. A dam burst and released a wall of sludge, said to be 20 m high, which overwhelmed the small village of Stava. The fluidised material consisted of the earth and rubble of which the dam had been built and large quantities of deposit previously lying at the bed of the lake.

The dangers of mud avalanches are increasing for several reasons. Not only are more people getting in the way, but they are removing natural defences. As populations increase and pressures on land grow, developing countries are removing stabilising vegetation. The risks are being increased by permitting overgrazing, by destroying the previous land cover for crops and by cutting timber without control. Air pollution is a significant factor in killing trees. It has justly been said that 'debris flows may be natural, but they are not averse to a helping hand'.

9.4.1 Instabilities in debris and mud flows

We have seen how debris avalanches can flow like a liquid and that the ability to carry big boulders is the crucial characteristic of true debris flow. Field observations from a variety of sources show that debris flows in gently graded channels tend to a fairly uniform thickness, but a mode of instability appears under some conditions of speed, density and slope, these flows showing a pulsing nature with a series of large waves or surges.

Figure 9.8 Debris flow of muddy sand developed from the cone of sediment from a gravel washing plant. (a) An active flow. (b) The thinner 'dead' flow. (Courtesy of J.R.L. Allen.)

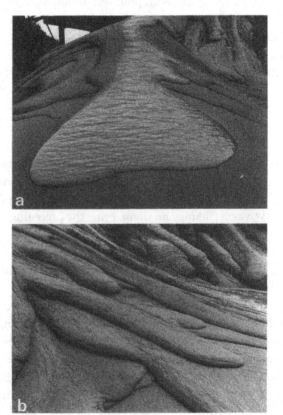

The loss of strength which accompanies the remoulding of muddy sediments and introduces a mudflow may, in some cases, be reversible. Materials which behave in this way are called *thixotropic*; some examples are emulsion paint, strong suspensions of cornflower or custard powder and many aqueous clay mineral dispersions. A thixotropic fluid has been used to demonstrate a mode of instability in laboratory suspension flow; figure 9.9 shows a very thick suspension of custard powder in water moving down a slope of 20° in a tank of fresh water.

Mudflows generated by heavy rainfall are rather common in the mountainous region of SW China and frequently cause catastrophic damage to the local villages and fields. Some of the largest and most hazardous mudflows occur in the Jiang-jia Ravine, a district which has been carefully studied by scientists from Sichuan Province (Li *et al.*, 1983).

One important feature which has been discovered from observations at the research station in the ravine has been named the 'bursting' process. This type of mudflow usually bursts in the form of successive waves, examples of which are illustrated in figure 9.10. The wave trains occur from time to time and there is little or no flow between successive groups of waves. The flow may cease altogether for a period of 20 or 30 minutes. This period of silence ends with a distant sound like thunder marking the next convulsion, and the surges of the mudflow rush down one after another until the episode is ended.

This kind of behaviour depends on the physical nature of the material and

Figure 9.9 Instability on the surface of a laboratory model of a debris flow. The material used is a very thick suspension of custard powder.

the rheological features of the fluid flow; the field work suggests that the transport of large solids causes the pulsing.

9.5 Glaciers

Like avalanches, glaciers are gravity currents flowing down mountain slopes; but as glaciers are made of ice formed from compacted snow their dynamics are entirely different (Paterson, 1981). Like soft putty, glaciers move very slowly through a combination of two processes, sliding on the bed and internal deformation. Figure 9.11 is a typical velocity profile through a glacier and shows both the basal sliding and the deformation, or 'creep'.

Glaciers can be subdivided into warm or cold varieties according to their internal temperature profiles. The thermal regime has a profound effect on glacier movement, affecting erosion, transport and deposition. The temperature throughout warm glaciers, characteristic of most mountain glaciers, is close to the melting point of ice, so they can contain water, which generally makes them more mobile. Cold glaciers, which have much lower temperatures, are found in the polar regions and are relatively sluggish.

9.5.1 The glacial system

The overall rate of flow of a glacier is determined by the balance between accumulation of snow being compacted into glacier ice and the loss of ablation, a combination of evaporation and melting. Figure 9.12 is a profile through a glacier showing how in the upper regions the snow accumulates until an equilibrium line is reached in which the net gain of glacier ice is balanced by the

Figure 9.10 Advancing surge of mudflow in the Jiang-jia Ravine. (Courtesy of Sichuan Geographical Institute.)

steady losses by ablation. Further down-glacier there is a net loss throughout the ablation area.

The flow lines in the figure show how material deposited at the upper surface becomes buried and later emerges at the surface at the lower end.

If as much ice is formed in the accumulation area as is lost in the ablation area, then the glacier 'snout' remains in the same position.

Speeds vary from a metre a year up to 4000 at the centre in exceptional cases, and the sliding speed usually accounts for about one half of the total. As figure 9.13 shows, when a glacier moves on steep slopes the flow rates increase and it usually gets thinner; when it decelerates the glacier ice is compressed and thickened, forming crevices and icefalls where the ice surface is broken up into small blocks. These features can be seen in the view in figure 9.14 of a typical long and narrow valley glacier in the Mont Blanc range.

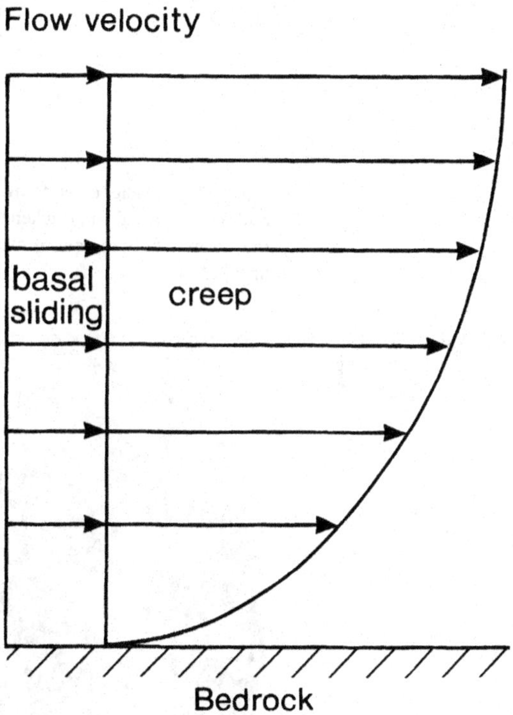

Figure 9.11 How a typical glacier flow varies with height. The movement is a combination of sliding over the ground and deformation.

Flow velocity

basal
sliding

creep

Bedrock

Figure 9.12 Profile through a glacier showing the accumulation and ablation areas. The flow lines show how material becomes buried and later emerges at the surface.

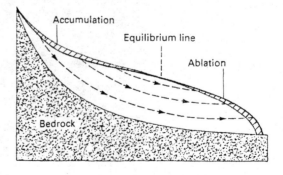

Accumulation

Equilibrium line

Ablation

Bedrock

Glaciers sometimes show unexpected increases in speed, or 'surges', when the snout may advance up to 60 or 70 metres per day, causing damage by damming lakes or blocking roads (Hambury & Alean, 1992). Various explanations have been suggested for these 'Galloping Glaciers', relying on mechanisms which

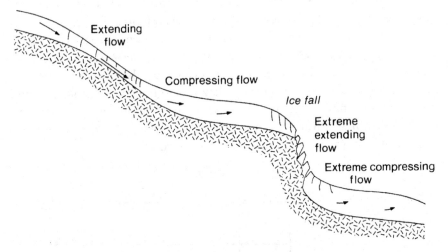

Figure 9.13 A glacier extends and thins where the speed increases, and is compressed and thickened where it slows down. Crevasses may form in both regions of extension and compression. (After Sharp, 1988.)

Figure 9.14 A glacier in the Mont Blanc range, showing crevasses and a typical icefall.

reduce friction by partly separating the ice from the bedrock. A fast-moving surge front then propagates downglacier as a wave which reaches the snout and initiates a sudden advance.

9.5.2 Glacier erosion and deposition

Glaciers are powerful agents of erosion at their bed, walls and sides and shape the landscape by scooping out and transporting rocks. The two dominant processes are abrasion (like sandpapering) and plucking, in which rocks are loosened and entrained into the flow. The U-shaped valleys produced by glaciers are different from river valleys; some of these in fjords are as much as 1000 m deep.

On the geological time-scale vast advances of the polar icecaps occurred in the Ice Ages, profoundly influencing the form of the earth's surface. During the periodic advance of the glaciers in the Pleistocene epoch most of eastern North America was covered far south of the Great Lakes; in Europe ice extended into Germany and Poland and Britain supported its own ice cap. The effects of glaciation can be clearly seen today in all these areas.

Glaciers have a tremendous capacity for transporting material in and on and beneath the ice; the process is slow but everything must in time be laid down. The deposition of such materials is illustrated in figure 9.15, in which they eventually approach the ice margin and are finally deposited as part of the moraine. If the glacier recedes the particles will be left in the terminal moraine.

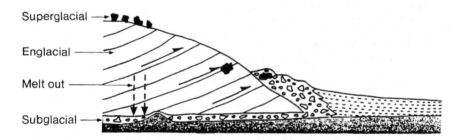

Figure 9.15 Processes in the transport and deposit of glacial material.

Bibliography

Beaty, C.B. (1963). Origin of alluvial fans, White Mountains California and Nevada. *Ann. Ass. Am. Geogr.*, **53**, 516–35.

Davies, T.R.H. (1982). Spreading of rock avalanche debris by mechanical fluidisation. *Rock Mech.*, **15**, 9–24.

Hambury, M. & Alean, J. (1992). *Glaciers*. Cambridge: Cambridge University Press.

Hopfinger, E.J. (1983). Snow avalanche motion and related phenomena. *Ann. Rev. Fluid. Mech.*, **15**, 47–76.

Hopfinger, E.J. & Tochon-Danguy, J.C. (1977). A model study of powder snow avalanches. *J. Glaciol.*, **19**, 343–356.

Li Jian, Yuan Jianmo, Bi Cheng & Luo Defu (1983). The main features of the mudflow in Jiang-Jia Ravine. *Z. Geomorph. N.F.*, **27**, 325–41.

Melosh, H.J. (1980). Acoustic fluidisation: a new geologic process. *J. Geophys. Res.*, **84**, 7513–20.

Paterson, W.S.B. (1981). *The Physics of Glaciers*. Oxford: Pergamon Press.

Salm, B. (1966). Contribution to avalanche dynamics. *Symp. Int. sur les Aspects Scientifique des Avalanches de Niege. Davos Switzerland. AIHS Publ.*, **69**, 199–214.

Salm, B. (1982). Mechanical problems of snow. *Rev. Geophys. Space Phys.*, **20**, 1–19.

Scheiwiller, T. (1986). Dynamics of powder-snow avalanches. *Report No. 81, Mitteilungen der versuchsanstalt für Wasserbau Hydrologie und Glaziologie*. Zurich, 115 pp.

Sharp, R.P. (1988). *Living Ice*. Cambridge: Cambridge University Press, 225 pp.

10   Volcanic gravity currents

Several types of gravity current are associated with volcanic eruptions. The emergence of material from the magma, the molten rock beneath the ground, forms currents whose nature depends on the viscosity of the magma and on the quantity of dissolved gas contained under pressure.

At one end of the range are liquid currents of basaltic lava, comparatively fluid molten rock at temperatures of 1200 °C which emerges quietly and flows as a glowing stream down the mountainside. Some laval flows of very viscous molten rock rich in silica form steep-sided plugs and domes, and can take months to move a few hundred metres.

If the magma is viscous and contains large quantities of gas, the nature of the eruption may be quite different. A vertical explosive eruption column of fragmented material (pyroclasts) may be projected hundreds of metres up into the air. Some material forms buoyant plumes which flow as intrusions at their neutral buoyancy level in the atmosphere.

Other material falls back under gravity and forms a pyroclastic gravity current. Such a current has sometimes been called a *nuée ardente*, or fiery cloud. Pyroclastic gravity currents from the largest eruptions may still have velocities of up to 100 m s^{-1} at distances of tens of kilometres from the source (Francis, 1993), so they form a serious hazard to people near volcanoes.

10.1 Basaltic lava streams

Basalt is the commonest kind of lava and when erupted is a runny liquid, glowing reddish-yellow at a temperature of about 1200 °C. Lavas are comparatively harmless, as one can see them coming and usually get out of the way quickly enough. Figure 10.1 shows a thin lava flow on the SW slope of Surtsey, Iceland, in 1963. It can be seen flowing down to the sea from the lava pond higher up the mountain.

As it cools, the lava gradually stiffens and develops a chilled skin which affects the nature of the flow. This skin can become wrinkled by the flow movement, giving a characteristic ropey appearance, or the skin can be torn into ragged fragments to produce rough, clinkery lava. As the material continues to stiffen, banks of solid material, levees, form at the sides. The whole flow may even end up confined to the inside of a solidified lava tunnel.

In the later stages, solid lumps of lava pile up in front of the advancing nose, which rolls forward and engulfs the material, rolling it up like the tracks of a caterpillar tractor. Later still there is no longer any fluid flow over the mass of rubble, which is then continuous piled up and pushed forward by the fluid behind it until at last the whole system comes to rest.

10.1.1 Very viscous lava flows

Silica-rich rock may still flow as a very viscous fluid. Such flows move at the speed of a glacier and can take months to cover a few hundred metres.

A good example of a lava dome formed by very viscous, slow-moving lava was studied in 1979 in the crater of Soufrière Volcano, St Vincent (Huppert *et al.*, 1982). The extruded lava formed a dome 130 metres high and 870 metres in diameter. The map of the crater in figure 10.2 shows how, in a period of three months, the leading edge of this actively growing viscous gravity current moved about 200 metres. Views of the Soufrière extrusion, taken on 4 August 1979, are shown in figure 10.3. A theory for such viscous flows was tested in the laboratory (see Chapter 15) and gave good results when applied to the Soufrière dome.

10.2 Pyroclastic plumes

In an explosive eruption where the magma is especially hot, and therefore of low viscosity, and contains large amounts of dissolved gas, the pyroclastic material reaches the surface at a high temperature and often with considerable momentum. A good picture of the process involved is of a fizzy drink being shot from a bottle which has been shaken. When the stopper is removed, fluid is

Figure 10.1 Lava flow on Surtsey, Iceland, in 1963. The lava flows from a lava pond down to the sea. (Photograph by Solarfilma, Reykjavik.)

blown with great force through the opening, and as the pressure is reduced large quantities of gas are released, forming a turbulent jet of material, and breaking up the fluid into small drops.

This volcanic material mixes with the air and heats it to provide buoyancy. The plume rises to its neutral buoyancy level then intrudes laterally as a gravity current (Sparks, 1986).

Material is then sedimented from this airborne gravity current, falls to the land surface and drapes it to form 'fall deposits'. This process is illustrated in figure 10.4a, which shows the result of deep-level explosive de-gassing.

10.3 Pyroclastic gravity currents

At the other end of the mobility scale from lava flows are the pyroclastic gravity currents shown in figure 10.4b, resulting from shallow-layer explosive de-gassing. Some of the very different forms taken by these gravity currents can be associated with the early stage of their formation, for example, the vertical column may collapse vertically or explosions can direct the volcanic material sideways. This occurred at Mount St Helens.

The first pyroclastic gravity currents to be studied scientifically were from Mount Pelée on Martinique in the Caribbean. The town of St Pierre was overwhelmed by a directed blast in 1902 and all but two of its inhabitants were killed (Francis, 1993). After some more eruptions of the same volcano, Perret (1937) studied hundreds of pyroclastic gravity currents and took a remarkable series of photographs. Grouping them all under the expressive name of 'Nuees Ardentes' Perret was fascinated by the appearance of the advancing front, and wrote, '. . . convolutions grew out of a rolling mass of incandescent material,

Figure 10.2 Map of the Soufrière crater, showing outlines of the lava extrusion at different times. The numbers refer to the observation points. (Courtesy of H.E. Huppert.)

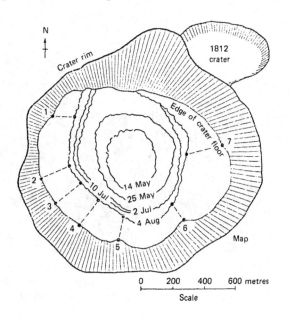

advancing with an indescribably curious rolling and puffing movement which at the immediate front takes the form of forward-springing jets, suggesting charging lions' – a graphic description of the many gravity current fronts illustrated in this book.

The name '*nuées ardentes*' is now known to include a large range of different types of pyroclastic gravity current in the atmosphere, in which clouds of ash material are generated that obscure much of what is going on in the clouds.

Figure 10.3 Views of the Soufrière lava extrusion on 4 August 1979. (Courtesy of H.E. Huppert.)

The currents beneath the clouds vary from the dilute ones, commonly known as 'surges', which have less fragmented material relative to gas volume, to those where the fragments are transported primarily by turbulent suspension. (Valentine, 1987).

Concentrated currents with a higher solid-to-gas ratio, commonly known as 'flows', may have bulk densities of 400–1500 kg m^{-3}. In these flows the fragments form a dispersion (i. e. turbulence is not an important mechanism of particle support; particles are supported by fluidisation processes, etc.) with a distinct interface between the concentrated flow and its more dilute ($\rho \leq 2$ kg m^{-3}) cloud above.

A comparatively recent photograph of this effect is shown in figure 10.5, which was taken at Mount St Helens in 1980. Although little of the complicated internal structure can be deduced from this view, it does show the raised nose and lobe-and-cleft structure of a typical gravity current.

Two stages in the life-history of a small pyroclastic gravity current are shown in the photographs of figure 10.6. The first photograph was taken only a few seconds after the start of an eruption at Ngaurahoe, New Zealand, and it

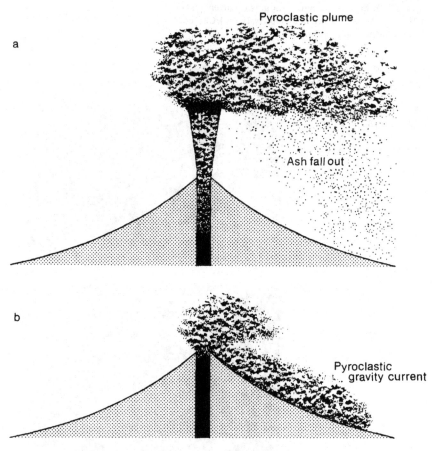

Figure 10.4 The contrast between (a) pyroclastic plumes and (b) pyroclastic gravity currents.

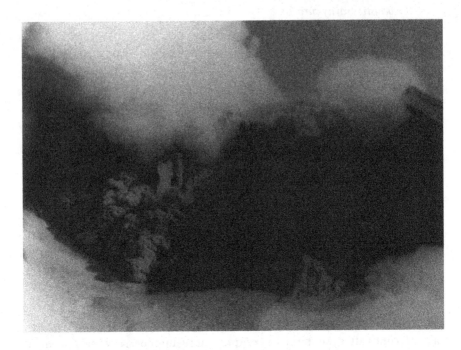

Figure 10.5 Formation of a pyroclastic gravity current on Mount
St. Helens, 7 August, 1980. (Photograph by Harry J. Glicken,
courtesy of US Geological Survey.)

Figure 10.6 Formation of
pyroclastic gravity current on
Ngaurahoe, New Zealand. (a)
A few seconds after start of
eruption, (b) About ninety
seconds later. (Courtesy of
G.T. Hancox, New Zealand
Geological Survey.)

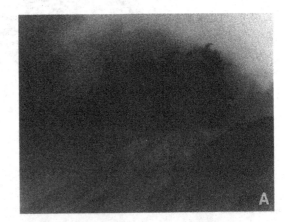

shows that the ejected material has already formed a large cloud. There are slight signs beneath this cloud of descending material. The second view was taken 90 seconds after the start, by which time a pyroclastic gravity current has formed from the material descending at the peak and its leading edge has already moved as far as the base of the mountain.

Mechanisms of fluidisation believed to be responsible for the rapid flow of other types of avalanche have already been discussed, both in relation to laboratory experiments and to avalanches of snow and rock. An additional process which may be important in the mechanics of these pyroclastic flows is the release of large quantities of gas which may come out of solution from the hot material. Some measurements of the actual speeds were made from an aeroplane flying over the eruptions of Mount Augustine Volcano, Alaska, in 1976 (Stith *et al.*, 1977). Along the initial 1:3 slope from 1200 m to 240 m an average speed of 50 m s^{-1} was attained. The speed reduced to 21 m s^{-1} down the subsequent 1:7 gradient, and the flow finally entered the sea, 6 km from the volcanic cone, after a time of 6.7 minutes.

From measurements of the deposits left behind these currents, deductions have been made about their internal progress. The transported mass may become more dense near the ground, and a buoyant cloud may separate (Hoblitt, 1986; Wilson, 1986).

The entrance of hot pyroclastic flows into water must be a frequent event on island volcanoes. There is evidence that these currents can move under water without losing their essential character and their deposits have been traced to over 13 km offshore at a water depth of 1800 m (Sparks *et al.*, 1980). Only flows denser than water are capable of a smooth transition into water, and conditions favourable for the passage into deep water include steep slopes and large flow rates. Laboratory experiments (Hampton, 1972) suggest that the shearing which occurs at the upper surface of the flow front generates a turbulent zone of mixing, which may continue as a turbidity current after the debris flow has come to rest.

10.4 Mud flows or lahars

Mud flows, or 'lahars' are common on many volcanoes but their importance is not always appreciated by the casually interested as they tend to get lost in press reports in vague description of 'eruptions'. Volcanic eruptions may trigger off mudflows both indirectly and directly. An eruption ejecting ash high into the atmosphere may propagate a tropical rainstorm, which in turn may initiate the kind of cold mudflow described in Chapter 9. Mud flows can also result when an eruption blasts through a crater lake, expelling the water and a great deal of volcanic debris at the same time. The lake water may be boiling hot, so that the mudflow down the mountain is a lethal mixture of scalding water, mud and boulders.

Figure 10.7 shows a phenomenon which is common in Iceland, where several active volcanoes lie beneath large ice sheets. Enormous quantities of water are melted beneath the ice-cap during a volcanic eruption and burst out, running down the mountain side. The resulting flow of slurry is known as a jökulhlaup.

Lahars have claimed more lives worldwide over the past 200 years than any other type of avalanche. One of the worst disasters was caused by the volcanic eruption of Nevado del Ruiz in Colombia on 19 November 1985, which killed 25 000 people. This eruption produced pyroclastic surges which generated four major lahars; these stripped vegetation and soil from the canyon walls to heights over 50 m and travelled a total distance of 60 km at a mean speed of 10 m s^{-1} (Lowe *et al.*, 1986). In the town of Armero the lahars overtopped the banks of the river channel and followed the old river course directly into the town, where most of the casualties were caused.

Analysis has been carried out of the remains of the debris flow in the canyon 2.5 km from Armero where the lahars stripped vegetation and soil from the canyon walls and deposited a layer of mud. Figure 10.8 shows the superelevation of the flow along the right bank as it rushed round the corner. This superelevation has been used to calculate directly the speed, w, of the debris flows

$$w = (g \cos A \tan B \cdot R)^{\frac{1}{2}}$$

where g is the acceleration due to gravity, A the slope of the stream bed, B the slope of the banked flow surface (deduced from differential levels of scour on the opposite sides of the river) and R is the radius of curvature of the bend. Around this bend, $g = 9.8 \text{ m s}^{-1}$, B is 7.5 degrees and R is 110 m, giving a mean front velocity of 11.9 m s^{-1}. This flow was about 45 m deep at mid-channel and had a cross-sectional area of nearly 4000 m^2.

One group of investigators (Lowe *et al.*, 1986) has pointed out that the large volume of glacial ice on the summit of the mountain, the abundance of debris around the summit and the precipitious gorges leading down to river valleys collectively represent ideal conditions for the formation and long-distance movement of lahars.

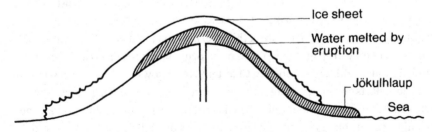

Figure 10.7 A jökulhlaup, a flow from a sub-glacial eruption.

10.5 Eruption of Mount St Helens

Descriptions of the explosive eruptions at Mount St Helens, Washington in 1980 illustrate well some of the processes mentioned above, namely pyroclastic gravity currents, rock avalanches and mud flows (Lipman & Mullineaux, 1981).

After two months of minor earthquake activity, in the morning of 18 May, 1980, there was a regional earthquake and the entire north side of the volcano began to separate from the main mass. A diagramatic illustration is shown in figure 10.9.

The detachment of the giant rockslide triggered explosions releasing a laterally directed pyroclastic surge from the mountain side. This material swept forward at over 90 m s^{-1}, overriding the rock avalanche, and devastated a sector reaching 20 km from the mountain summit. In an inner zone of 10 km radius virtually everything was destroyed, and beyond this radius all trees were blown down.

The progress of the pyroclastic gravity current seemed almost independent of the contours of the ground, but the advance of the rock and mud flows was somewhat different. The avalanching rock from the north flank acquired melted snow and ice blocks from the former glaciers of the volcano. Travelling at 80 km h^{-1}, this debris flow was diverted to the left by a ridge, displaced the water of Spirit Lake and flowed down the valley of the Toute River. For over 18 km this debris flow swept down the valley, causing damage, including destruction of

Figure 10.8 The superelevation of the lahar flowing down this canyon, near Armero, Colombia has stripped the right bank of all vegetation. A man (arrowed) shows the scale. (Courtesy of *Nature*, photograph by Donald R. Lowe, Department of Geology, Louisiana State University.)

bridges and houses. The deposit left in the upper reaches of the Toute River had an average thickness of 66 m.

The total number of people killed was just over 100, a comparatively small number for such a vigorous eruption owing to the district being only sparsely populated.

10.6 The Lake Nyos gas disaster

On 21 August 1986 a large volume of toxic gas was released from beneath and within Lake Nyos in the mountains of Cameroon. A gravity current consisting of an aerosol of toxic gases and water droplets swept down the valleys to the north, killing 1700 people.

The conclusions of a group of British investigators, after visiting the site (Freeth & Lay, 1986) are as follows. The waters of Lake Nyos were already saturated with CO_2 of volcanic origin when a pulse of volcanic gas, mainly of CO_2 but with some H_2S, was released in the lake above a volcanic vent. The rising stream of bubbles brought up bottom water containing CO_2 dissolved under pressure. As the pressure was released the gas came out of solution extremely vigorously, as when released in a fizzy drink, and increased the flow of water to the surface.

At the surface the release of gas transformed the water into a fine mist and this aerosol of water and heavy gases swept down the valleys to the north of the lake. It was estimated that about 200 000 tonnes of water were lost from the lake,

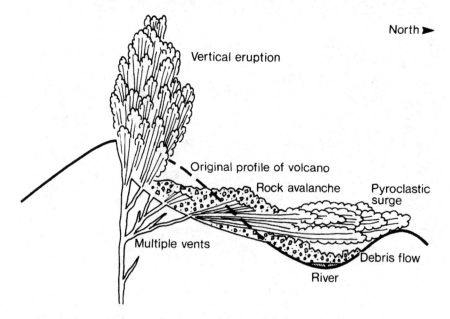

Figure 10.9 Section through Mount St Helens, showing the effect of the sideways blast in the eruption of 18 May, 1980.

together with 6000 tonnes of gas, which at atmospheric temperature and pressure would have a volume of about 3 million cubic metres.

Whilst this disaster is included in the section on volcanoes, the apparently stealthy approach of the aerosol current could classify it with the gas spills described in section 6.2.

10.7 Volcanic action on the ocean bed

Three quarters of the annual volcanic production occurs under water, mostly along the extensive mid-ocean ridge system (Fisher & Schminke, 1984).

Hydrothermal vents are found on the sea bed at great depths in both the Atlantic and Pacific Oceans. These emit buoyant plumes of hot water at over 300 °C, which convey many minerals in suspension. The commonest ones produce a black precipitate of sulphides and are called 'black smokers'.

The sites, which may be 3000 metres below the surface, act as a host on the sea bed for unusual communities of life, which are dependent on the high temperatures. These arise around edifices which form at the vents, mostly close to the white rather than the black chimneys, as shown in figure 10.10.

The high temperatures lead to important circulations in the ocean. The ascending plumes mix with their surroundings and are found to rise 100–500 m above source, eventually spreading out as an intrusive gravity current, as shown in figure 10.11.

Figure 10.10 Shrimps near the ocean bed, feeding on bacteria at a black smoker, seen to the right. (Courtesy of H. Elderfield.)

As well as these small continuous sources there exist much larger-scale episodes, which give rise to the so-called 'megaplumes'. These ascend over 1000 m and the horizontal currents formed by them have been traced to distances of over 100 km. Figure 10.12 (from Baker & Massoth, 1987) shows the plan view of such a megaplume, measured from the temperature anomaly at the 1600 m surface.

These plumes carry dissolved chemicals which play an important part in the total ocean budget. They also carry many particulates, whose fall-out may not be very important in the dynamics of the flow, but which are significant in their long-term deposits on the ocean bed.

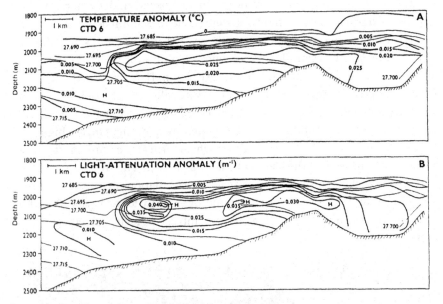

Figure 10.11 Intrusive gravity current which emanates from a number of axial valley vents shown from transects of (A) temperature and (B) light attenuation (Courtesy of E.T. Baker.)

Figure 10.12 Plan view of megaplume, showing temperature anomoly contours on 1600 m surface. (Courtesy of E.T. Baker.)

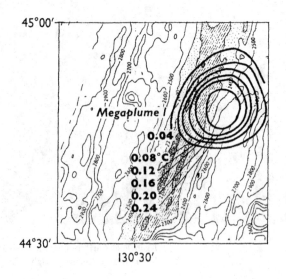

Bibliography

Baker, E.T. & Massoth, G.J. (1987). Characteristics of hydromechanical plumes from two vent fields on the Juan de Fuca Ridge, northeast Pacific Ocean. *Earth Planet. Sci. Lett.*, **85**, 59–73.

Fisher, R.V. & Schmincke, H.O. (1984). *Pyroclastic Rocks*. Berlin: Springer–Verlag, 472 pp.

Francis, P.W. (1993). *Volcanoes: A Planetary Perspective*. Oxford: Clarendon Press, 443 pp.

Freeth, S.J. & Lay, R.L.F. (1986). The Lake Nyos disaster. *Nature*, **325**, 104–5.

Hampton, M.A. (1972). The role of subaqueous debris flow in generating turbidity currents. *J. Sediment. Petrol.*, **42**, 775–93.

Hoblitt, R.P. (1986). Observations of the Eruptions of July 22 and August 7, 1980, at Mount St. Helens, Washington. *US Geol. Survey Prof. Pap.* 1335.

Huppert, H.E., Shepherd, J.B., Sigurdsson, H. & Sparks, R.S.J. (1982). On lava growth with application to the 1979 lava extrusion of the Soufrière of St. Vincent. *J. Volcan. Geo. Res.*, **14**, 199–222.

Lipman, P.W. & Mullineaux, S.R. (1981). The 1980 eruptions of Mount St. Helens, Washington. *US Geol. Surv. Prof. Pap.* 1250.

Lowe, D.R., Williams, S.N., Leigh, H., Conner, C.B., Gemmell, J.B. & Stoiber, R.E. (1986). Lahars initiated by the 13 November 1985 eruption of Nevado del Ruiz, Columbia. *Nature*, **324**, 51–3.

Perret, F.A. (1937). The eruption of Mt. Pelee 1929–1932. *Carnegie Inst. Washington. Pub.* **458**. 126 pp.

Sparks, R.S.J. (1986). The dimensions and dynamics of volcanic eruptions. *Bull. Volcano.*, **48**, 3–15.

Sparks, R.S.J., Sigurdsson, H. & Carey, S.N. (1980). The entrance of pyroclastic flows into the sea. *J. Volcanol. Geotherm. Res.*, **7**, 87–96.

Stith, J.L., Hobbs, P.V. & Radke L.F. (1977). Observations of a nuée ardente from the St. Augustine Volcano. *Geophys. Res. Lett.*, **4**, 259–62.

Valentine, G.A. (1987). Stratified flow in pyroclastic surges. *Bull. Volcanol.*, **49**, 616–30.

Wilson, C.J.N. (1986). Pyroclastic flows and ignimbrites. *Sci. Prog., Oxf.*, **70**, 171–207.

11   The anatomy of a gravity current

11.1 The front

The leading edge of a gravity current forms a typical frontal zone, i.e., although intense mixing is present, a sharp dividing line is maintained between the two fluids. A characteristic 'head' which is deeper than the following flow is usually formed at the front. This raised head is a zone of breaking waves and intense mixing and plays an important part in the control of the current which follows. In a gravity current moving horizontally the head remains quasi-steady, but in one flowing down an incline the relative size of the head increases with the angle of the slope.

A front advancing over a rigid flat surface usually has a foremost point or 'nose' which is raised above the following flow. Both the raised nose and the zone of intense mixing can be seen clearly in the photograph of the atmospheric gravity current, or 'haboob', illustrated in figure 1.1.

It is not possible to give a unique shape for the outline of the head of a gravity current, as even in flows into calm surroundings the excess head height above the following flow varies with the fraction of the total depth of the fluid occupied by the current. Apart from viscous effects, the head profile is also strongly modified by opposing or following ambient flows, and by turbulence in the surroundings.

The influence of viscosity on the behaviour of gravity currents has been examined in the laboratory, and it has been shown that viscosity affects the profile of the head and the rate of advance of the current. The results of some early experiments (Schmidt, 1911) showed this well, as illustrated in figure 11.1. A series of gravity currents of increasing temperature differences were used to model the advance of a cold squall in the atmosphere. These shadow-pictures show gravity currents with a temperature difference increasing from a very small value in (a) to 35 °C in (f), corresponding to a density difference of about 1%. As the temperature difference increases, and with it the speed of the gravity current, the shape of the head of the current changes. In the flow (a) at small temperature difference, viscous forces predominate over the forces due to buoyancy; as a result the head is small and there is very little mixing apparent. As the density difference increases the flow approaches the limiting profile in (f) in which eddies are formed and can be seen streaming back in the upper surface of the head.

The values of Reynolds number, UH/v, in the experiments shown in figure 11.1 increases from less than 10 in (a) to greater than 1000 in (f). The final view (f) shows a profile which has been shown (Simpson & Britter, 1979) to be independent of Reynolds number, that is, to be typical of all flows with Re greater than 1000. In all these flows a turbulent mixing pattern is seen streaming back in the upper surface of the head, and the height of the foremost point of the nose is about 1/10 of the total height of the head. The Reynolds number of a typical thunderstorm outflow in the atmosphere is about 1 000 000, and many field observations have shown the structure to be similar to that seen in the laboratory with Re greater than 1000.

The features of the head of a gravity current can be seen in the view of a laboratory experiment shown in figure 11.2. This shows the frontal region of

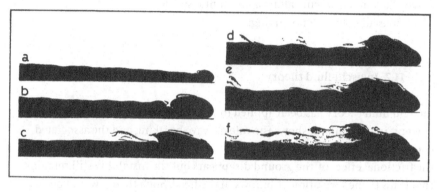

Figure 11.1 Shadow pictures showing profiles of the head of a gravity current. The temperature difference increases from very small in (a) to 35 °C in (f). From Schmidt (1911).

Figure 11.2 The frontal region of a gravity current of salt water advancing along the floor of a fresh-water tank.

a 1% saline solution advancing along the floor of a parallel-sided channel of fresh water 30 cm wide and 20 cm deep. The front is made visible by adding milk to the dense fluid.

The mixing processes seen both in the laboratory flows and in many environmental examples of gravity currents are complicated. Figure 11.3 shows the two main types of instability which are responsible for the mixing. These consist of (a) billows, which roll up in the region of velocity shear above the front of the dense fluid, and (b) a complex shifting pattern of lobes and clefts, which are formed by the influence of the ground on the lower part of the leading edge.

The next section considers the simplified front of a two-dimensional gravity current advancing along a horizontal surface, firstly in the simplest form of an inviscid flow with no mixing.

11.2 Inviscid-fluid theory

Inviscid-fluid theory has been applied to study aspects of a steady gravity current; in particular it shows the role of wave-breaking and the associated energy losses. In an inviscid fluid, in which viscous forces are completely absent, the frictional effect of the ground disappears but the instability still remains that leads to the formation of billows in the head. Such theory would be expected to give a useful approximation to the behaviour of the front of a gravity current, at least for flows with Reynolds number above a certain value.

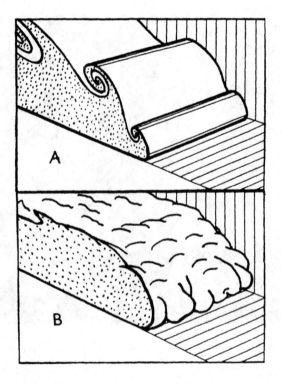

Figure 11.3 Two forms of instability at the front of a gravity current moving along the ground: (a) billows, (b) lobes and clefts.

Nevertheless, some of the details must be different from flows in real fluids, in which frictional forces certainly play a part. Such a theoretical 'inviscid current' cannot be realised in nature, but can be approached experimentally in certain ways.

The front of a frictionless gravity current has been analysed by Benjamin (1968) in terms of a 'cavity flow' displacing a fluid beneath it. If one end is removed from a long, closed water-filled channel, the water will begin to run out at the lower level, being replaced by air flowing in above it. Figure 11.4 shows the form of the air intrusion. The velocity U_2 is relative to the leading edge of the cavity.

Two equations involving the velocity and depth h_2 of the flowing layer can be obtained from continuity of mass and by the use of Bernouilli's equation applied along the interface. (Bernoulli's equation expresses the constancy of total energy along a streamline, where the sum of the three types of energy in the flow, pressure, potential and kinetic, remains constant.) Being frictionless, the 'flow force' (total pressure force plus the momentum flux per unit span) is also conserved, resulting in a solution

$$h_2 = H/2$$

Thus the only steady energy-conserving flow (implied by the use of Bernoulli's equation) is one in which the advancing layer fills half the channel. Flows in which $h > H/2$ are not possible, and if $h < H/2$, as found in most practical situations, the loss of energy at the front exceeds that available by wave radiation, so that 'breaking' must occur.

It can also be shown that the fractional depth $h/H = \phi$ say, plays an important part. The Froude number, $U/(g'h)^{\frac{1}{2}}$, based on the velocity U of the front, is $2^{-\frac{1}{2}}$ at $\phi = 0.5$, and equals about 1 at $\phi = 0.2$, increasing to $2^{\frac{1}{2}}$ as ϕ tends to 0.

In subsequent treatment of the velocity of the head under 'deep' water (i.e. ϕ tends to 0), if the pressure downstream in the wake is taken to be hydrostatic so that the dynamic pressure $\frac{1}{2}\rho U^2$ at the stagnation point at the front boundary equals the difference between the hydrostatic pressure at the boundary far upstream and downstream, it follows that

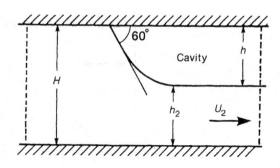

Figure 11.4 Theoretical treatment of a gravity current head. The fluid is displaced as it passes the cavity.

$$\tfrac{1}{2}\rho U^2 = g(\rho_1 - \rho_2)h$$

or

$$U = 2^{\tfrac{1}{2}}(g'h)^{\tfrac{1}{2}}$$

The same result had been previously obtained by Von Karman (1940), by applying the Bernoulli condition along the interface, but this is invalid in deep water since the interface must be dissipative and Bernoulli's equation cannot be applied.

The approximate shape of the interface has also been calculated, and the slope at the stagnation point on the ground can be shown to be 60 degrees.

11.3 Effect of mixing on gravity current

Most gravity currents in the environment have turbulent mixing at the front, but there may be only small frictional effects if there is no rigid boundary. An example of this is a large-scale gravity current of fresh water moving above salt water, such as a river flowing out over the sea. In this case there is considerable mixing at the interface between the fresh and salt water, and this plays an important part in the dynamics of the flow.

A semi-empirical analysis of this type of front has been carried out as a first step towards realistic modelling of the head of such gravity currents (Britter & Simpson, 1978). In this analysis the flow relative to the head is divided into three regions, as shown in figure 11.5. This represents a gravity current of dense fluid, miscible with the surrounding fluid, moving along a plane surface with no friction.

The bottom region, depth h_4, represents the flow of dense unmixed fluid into the gravity current front. The top region contains only the unmixed less dense fluid, depth h_2, passing above the head. The section, h_3, between these layers is the mixing region. This region has non-uniform velocity and concentration profiles, determined by experiment.

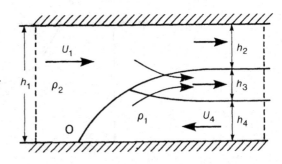

Figure 11.5 The flow relative to an inviscid gravity current with mixing. The bottom region is the flow of dense unmixed fluid into the front; above it is the mixed region of the collapsed billows.

The apparatus shown in figure 11.6 was used to examine the properties of the mixing at the head of a gravity current with effectively no friction at the floor. It consists of a parallel-sided channel of perspex with water pumped through it at a steady rate. The floor of the left part of the channel consists of a flexible conveyor-belt which can be moved at the same speed as the water. A steady supply of dense fluid is introduced, through a meter, from beneath at the right end of the tank.

For a fixed input of dense fluid of a given density it is possible to adjust the flow-and-floor speed to bring the dense fluid to rest just downstream of the conveyor-belt, as shown in figure 11.6a, forming a front at rest at the beginning of the fixed floor section.

It is important to realise the difference between this front and the familiar 'arrested saline wedge' of dense fluid brought to rest by a simple opposing flow along the ground. The wedge has little or no mixing and will be dealt with in more detail in section 11.5.1. As shown in figure 11.6, in case (a), with the floor moving at the same speed as the water flow, the front is brought to rest by a relative flow having a constant velocity profile, U. The effect is therefore the same

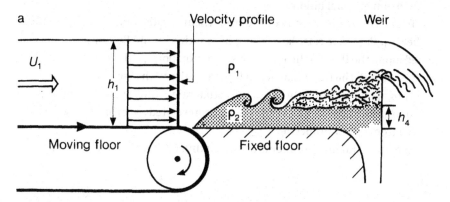

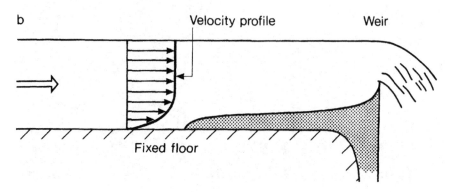

Figure 11.6 Apparatus used to maintain the head of a gravity current in a steady state. (a) The moving floor is used to modify the velocity profile in the approaching flow. (b) The whole of the floor is at rest.

as a front moving at velocity U into a calm fluid. However, in case (b), conditions are quite different; the wedge is brought to rest by a flow whose velocity is reduced by friction to zero close to the ground.

The nature of the imposed velocity profile has a critical effect on the instability and hence the mixing developing along the foremost interface. The result of the profile in figure 11.6b is to reduce the sharpness of the velocity profile in the flow along the interface; as a result there is an almost complete absence of instability at the front of the arrested wedge.

When the front is held stationary on the fixed floor just downstream of the end of the conveyor-belt section it appears as shown in figure 11.7. This photograph was obtained by a simple 'shadowgraph' technique, which is a simplified form of the 'schlieren' process, using deflection of the light by density gradients in the fluid under examination. A slide-projector is used at a distance of about 5 metres (or more) to project a shadow of the flow onto a transparent screen. It can be seen that the foremost point of the dense fluid is on the ground and it is found that any lobe-and-cleft instability (illustrated in figure 11.3) has completely disappeared. The mixing zone above the head can be seen to contain clear two-dimensional billows.

To determine the velocity of the front theoretically, values of the Froude number $U/(g'h_4)^{\frac{1}{2}}$ were calculated using the continuity and Bernoulli equations applied along the floor to the foremost stagnation point, O. This Froude number varied both with the fractional depth h_4/h_1 and also with $q=g'Q/U^3$, the non-dimensional mixing rate, where Q is the actual rate of mixing. Values of q needed to close the equations could either be measured experimentally or else deduced from billow properties.

Experimental measurements of the mixing rate were obtained in the

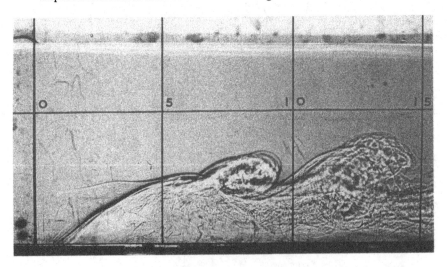

Figure 11.7 Shadowgraph of the head of a gravity current produced in the apparatus shown in figure 11.6a. The lobe-and-cleft instability has disappeared and the billows are two-dimensional.

apparatus of figure 11.6 by simply monitoring the rate of input of dense fluid needed to maintain the steady state of the front.

This apparatus also made it possible to investigate the properties of the billows forming at the front, and it was found that they had the properties of Kelvin–Helholtz (K–H) billows. This type of billow is associated with instability formed at the interface between two fluids of different density moving relative to each other. The instability can occur for certain ranges of values of the velocity change U per distance h, with density difference given by diluted gravity g'. The relevant dimensionless number here is the Richardson number Ri, which may be put in the form $(g'h)/U^2$. Kelvin–Helmholtz instability usually occurs for a value of Ri less than $\frac{1}{4}$.

Measurements of Kelvin–Helholtz instability have been carried out for large ranges of velocity and density shear by Thorpe (1973) and it appears that the breakdown size of the billows at a gravity current head agrees with that of K–H billows forming at a very low Richardson number (i.e. a very sharp interface, as exists at the foremost part of a gravity current). The height of a gravity current head is identical with the breakdown size of the K–H billows which are being formed there.

Using either of these methods of obtaining the non-dimensional mixing rate, the Froude number was found theoretically, and confirmed by experiment, to be close to 1 for fractional depth h_4/h_1 about one fifth of the total depth. For smaller fractional depths, down to about 0.05, the value of the Froude number rose to approximately 2.

11.4 Effect of friction

The outline of the front of a gravity current advancing along the ground has appeared in several of the environmental photographs, and shows a rather complicated system of billows, and also lobes and clefts. The shadowgraph in figure 11.8 gives us a look inside such a front in the laboratory. An almost unmixed dense saline flow can be seen entering the water tank on the right at the bottom, and the front of the gravity current is moving to the left along the floor of the tank. The foremost point is raised from the ground, and scattered Kelvin–Helmholtz billows can be seen forming above and collapsing to the right of the head.

The schematic diagram in figure 11.9 shows a simplified two-dimensional flow pattern relative to such a gravity current head. In such a flow, with no-slip conditions at the lower boundary due to friction at the stationary ground, the lowest streamlines in the flow relative to the head must be towards the rear. This means that the stagnation point 0 must be raised a small distance above the floor and in addition to the circulation in the upper part of the head there must be a smaller circulation in the reverse sense close to the ground. The fluid which has been shaded in the diagram is that which is destined to pass underneath the

Figure 11.8 Shadowgraph of a laboratory gravity current on a
horizontal flow. Salt solution enters on the right.

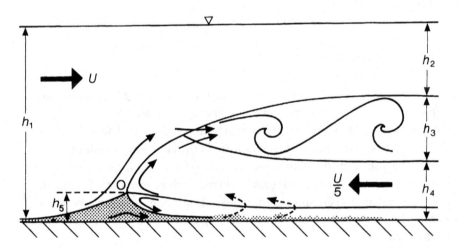

Figure 11.9 The flow relative to the head of a gravity current
advancing along a horizontal surface. The foremost point of the
gravity current, O, is raised above the floor, and the shaded fluid
passes beneath the head. There is a small reverse circulation close
to the ground.

nose of the current. This is less dense than the fluid above it and therefore is unstable. In the process of its ascent and forward movement it is responsible for the non-steady lobe and cleft structure.

In this type of flow along a solid boundary the billows which form at the leading edge are broken up in a complicated three-dimensional form. Nevertheless, by using special lighting it is still possible to distinguish perfectly formed Kelvin–Helmholtz instabilities (Simpson, 1969). An example appears in figure 11.10 which shows three successive views of a cross-section of a gravity current head taken at intervals of $\frac{1}{4}$ second. This was obtained by using a narrow sheet of light which illuminated a cross-section of a gravity current marked by a fluorescent dye.

The billows appear to be the main process in which the surrounding fluid is mixed into a gravity current. The fluxes in and out of the frontal region of a gravity current have been measured using hot-wire probes on a carriage moving with the front by Winant & Bratkovich (1977). It is found that the mass flux of heavy liquid into the front is of the order of 0.15 times the mass flux of the current itself.

To summarise, the three graphs in figure 11.11 show the variation of velocity of a gravity current with fractional depth. The speed is non-dimensionalised as the Froude number, $U/(g'h_4)^{\frac{1}{2}}$. (1) is the inviscid flow, (2) is the inviscid flow with mixing, and (3) is the flow along a horizontal surface.

11.4.1 Lobes and clefts

The complicated shifting pattern of lobes and clefts at the head of a gravity current moving along a horizontal surface is believed to be caused by gravitational instability of the less dense fluid which is overrun by the nose of the current.

The evolution of the lobe-and-cleft structure has been studied in experiments in which the plan view of a gravity current front was photographed every half second. Figure 11.12 shows a plan view of the results of one of these experiments, in which the dotted lines show the continuity of each cleft from its initial birth at point X. Clefts do not disappear but may absorb or be absorbed by their neighbours since all the lobes are either swelling or shrinking in width. There is a maximum size possible for a lobe, and when it reaches this, a new cleft forms in it. The figure shows how, in this rapidly shifting system, the total number of lobes and clefts can remain almost constant.

It was found that the breakdown size of the billows was twice the mean size and varied with Reynolds number in the range from Re=400 to about 4000. For greater values of Re the ratio of mean lobe size to total head height appeared to be nearly constant and had a value of about $\frac{1}{4}$.

The essential role played by the overrunning less dense fluid in causing the lobe-and-cleft instability has been confirmed by two simple laboratory experiments (Simpson, 1972), in which it was found possible to suppress the instability.

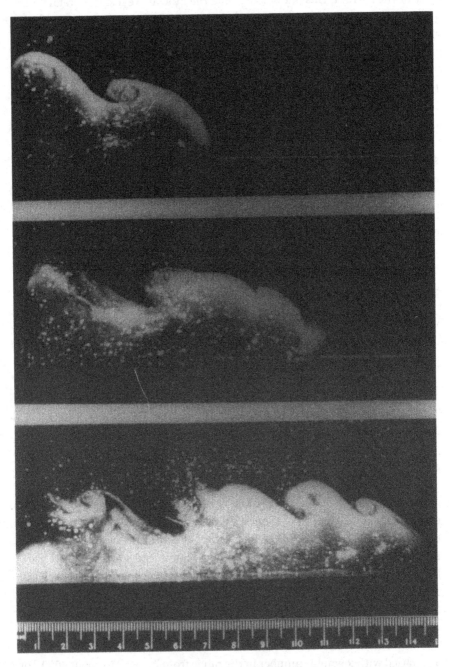

Figure 11.10 Three successive views of a gravity current head at intervals of $\frac{1}{4}$ second, in which some clear two-dimensional billows have been detected.

In the first experiment, the overrunning of light fluid was prevented by laying down a thin layer of dense fluid ahead of the current. This was deep enough to ensure that all the overrunning fluid was no longer lighter than the gravity current fluid; the result was the appearance of clear two-dimensional billows.

In the second experiment, a section of the floor was moved in the direction of the current, but only beneath the dense current itself. Ahead of the current the floor was stationary as usual. As the speed of the floor-section increased, the

Figure 11.11 The speed of the head of a gravity current, expressed as the Froude number, advancing in fluids of different depths (expressed as current depth/total depth). Curve (1) is the inviscid current with no mixing, (2) is an inviscid current with mixing, and (3) is a mixing current moving along the ground.

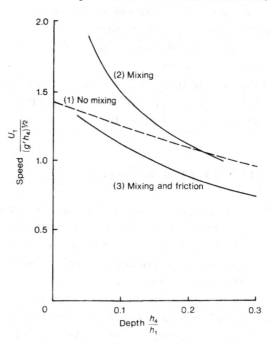

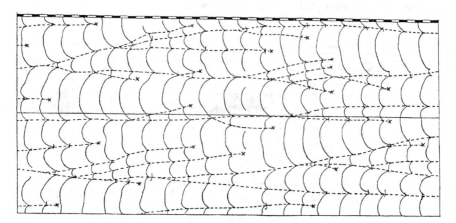

Figure 11.12 Successive views of the leading edge of a gravity current seen from above at intervals of half a second, showing the evolution of lobes and clefts. X marks the points where new clefts appear. The scale is in centimetres, $\Delta\rho/\rho$ is 1% and total water depth is 24 cm.

profile of the front became flatter, but also the foremost point steadily approached the ground. When the foremost point was actually on the ground the flow duly became two-dimensional and clear K–H billows appeared.

11.4.2 Height of the foremost point, or nose

If we now include the effect of friction on the boundary, the main difference from the inviscid model can be related to the height above the ground of the foremost point of the head, sometimes called the 'nose'.

Measurements of the height of the nose of a gravity current advancing along the ground into calm surroundings have been made in the laboratory and in the atmosphere and some results are shown in figure 11.13. Laboratory experimental results are shown for values of the Reynolds number, $U_1(h_3+h_4)/\nu$, from 10 to 10^5, and atmospheric results are at about $Re=10^8$. The results show a height of the foremost point of about $\frac{1}{8}$, apparently independent of Reynolds number for values greater than 10^3.

11.5 Head and tail ambient flows

The profile of the head of a gravity current and its rate of advance are very sensitive to any opposing or following flow in the environment. These head- and tail-flow effects on the front of a gravity current moving along a horizontal surface have been examined in a water channel with a moving floor of the type already described in section 11.4.

Using this apparatus, illustrated in figure 11.14, head flows, or 'head winds', are simulated by bringing to rest on the moving floor a front with a flow, U_1, greater than and opposing that of the floor, U_0. Tail winds are obtained by moving the floor faster than the opposing flow.

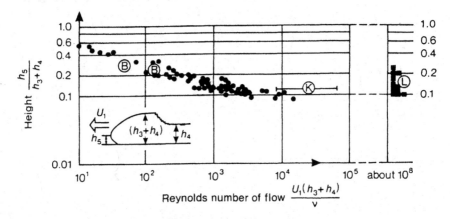

Figure 11.13 Values of the height of the foremost point of a gravity current head, measured for flows at different Reynolds number. B=Barr (1967), K=Keulegan (1957), L=Lawson (1971). The figures on the right are from atmospheric observations.

The experiments show that with a head wind the head profile is longer and the nose height lower than in the calm case already illustrated. With a tail wind the head is shorter and the nose height is larger. Figure 11.15 shows these three different forms.

A head or tail wind is found to change the speed of advance along the ground by about $\frac{3}{5}$ of the applied wind. Values of the overtaking speed relative to

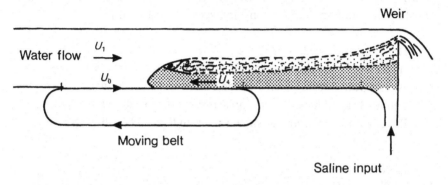

Figure 11.14 Schematic diagram of a water channel with a moving floor used to investigate the dynamics of gravity current head. The speed of the water flow U_1, the speed of the moving floor, U_0, and the input of dense fluid, Q, can all be controlled.

Figure 11.15 Shadowgraphs show the effect of head and tail winds on the form of a gravity current head. In the experimental tank the vertical lines are 10 cm apart. (a) Floor speed greater than flow speed (head wind). (b) Floor speed equal to flow speed (calm). (c) Floor speed less than flow speed (tail wind).

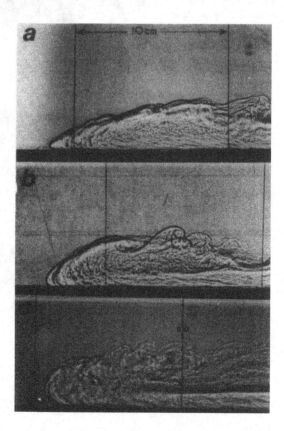

the front, U_4, were compared with U_1, the flow relative to the ground and the values of U_4/U_1 were all found to be close to 0.15, independent of the values of the head or tail flows (Simpson & Britter, 1980).

The value of the mixing rate $R_L = g'H_3/(\Delta U)^2$ was deduced to be 0.5 in all head- and tail-wind cases. This is greater than the values measured in the two-dimensional (inviscid) case, but close to that already found in the calm case along a horizontal floor. Thus, it appears that the shape of the head and mixed region adjusts to maintain this almost constant layer Richardson number.

11.5.1 Arrested wedges

At the mouth of many rivers, as the tide level increases, a gravity current of dense salt water moves upstream along the river bed. Eventually, the front is brought to rest and for some time an arrested saline wedge exists.

The laboratory experiment shown above in figure 11.6b displayed an

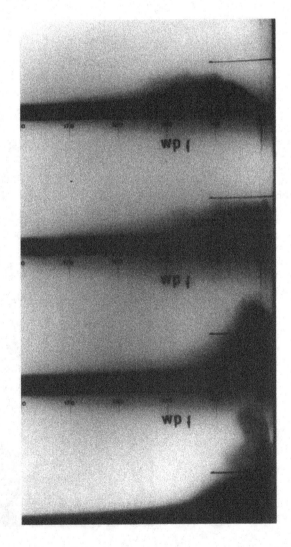

Figure 11.16 Four views at one-second intervals of an experimental collision of a gravity current front with a vertical wall, showing the 'splash' which moves up the wall. The original height of the dense fluid before its release is shown by the black horizontal line.

arrested wedge formed on a fixed floor by an opposing flow. Figure 11.6 shows the essential difference between this wedge and a gravity current moving into still fluid (or brought to rest by opposing flow and floor) (Figure 11.6a).

The difference was explained by the velocity profile near the ground in the flow which is meeting the wedge. This has the effect of reducing the instability along the leading edge of the dense fluid. In fact, the mixing along the top surface of an arrested wedge may be almost non-existent. Laboratory experiments by Riddell (1970) have shown that a considerable proportion of the interface of such a saline underflow can be very closely approximated to a straight line, and that no appreciable change in density occurs during the length of the underflow.

11.6 Gravity currents meeting obstacles

The problems of the effects of barriers on the behaviour of gravity currents have wide practical applications, for example on the control of accidental escapes of dense fluids or gases. Laboratory experiments have examined the effects of various kinds of obstacle, and have suggested methods of theoretical analysis and numerical simulation.

11.6.1 Gravity currents meeting a solid barrier

Figure 11.16 shows four sequential photographs from an experiment using saline flows in a water tank (Rottman *et al.*, 1985), in which a gravity current meets an 'infinitely high' barrier. A high splash of heavy fluid runs up the wall to about twice the original height of the released fluid (indicated by a dark horizontal line). This mass of fluid collapses and forms a raised disturbance moving away from the barrier, as shown in the last photograph.

A sketch of the flow under consideration is shown in figure 11.17. A simple solution of this can be obtained, assuming that h_1 is constant during the interaction. The result predicts a hydraulic jump propagating away from the wall after the interaction.

The nature of the return flow is more complicated than a simple reflection from a wall. The mixing which has been taking place at the head during the previous advance of the gravity current has left behind a stably stratified layer above the dense current, as described in section 11.3 and shown in figure 11.5. This property results in the hydraulic jump being manifested as primarily a smooth solitary wave moving back through the stable layer, progressing at a speed close to that of the original gravity current, but in the opposite direction.

Records made of the progress of the reflections of gravity currents from the end walls of laboratory tanks confirm that the first reflected disturbance travels back at a uniform speed not very different from that of the original gravity current. Figure 11.18 shows such a disturbance produced by reflecting a gravity current from the end wall of a tank. The appearance of the mass of fluid, seen moving to the left, is that of a solitary wave, and the presence of the dye patch, which was inserted at the end wall, makes it clear that some of the original mass is still being carried along.

Experiments have been made with barriers of less than twice the height of the current, in which some of the flow crosses the obstruction. These all show a similar splash and a reflected disturbance, but some of the fluid continues to flow on beyond the barrier and to form a gravity current on the other side. This is clearly shown in figure 11.19.

Work has been done on barriers of other forms. Dykes of different slopes have been employed and the dependence of spillage on the inclination of the dyke has been determined experimentally (Greenspan & Young, 1978).

11.6.2 Gravity currents flowing through a porous obstacle
The series of photographs in figure 11.20 shows the interaction of a gravity current with a porous barrier consisting of 40 evenly spaced wooden dowels.

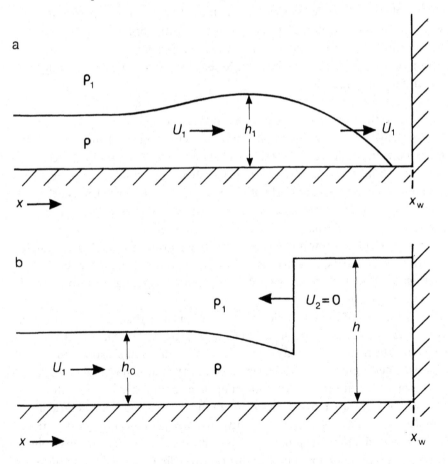

Figure 11.17 Schematic illustration of the interaction of a gravity current with a solid vertical wall. (a) Before the interaction, (b) after the interaction, with a hydraulic jump propagating away from the wall.

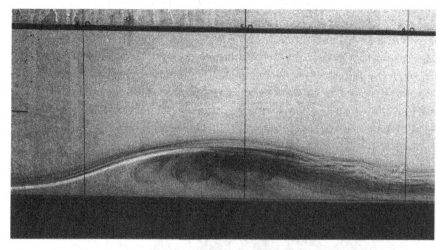

Figure 11.18 Shadowgraph of the disturbance produced by reflecting a gravity current from the end wall of a tank. The appearance is close to that of a solitary wave.

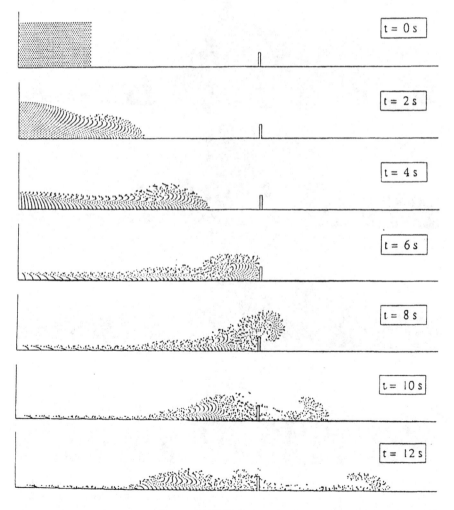

Figure 11.19 Numerical simulation of the flow above and through a fence. (Courtesy of K.P. Gröbelbauer and T.K. Fannelop.)

Firstly, the gravity current increases in depth as it encounters the barrier, and it can be seen that its height nearly doubles. Soon, however, the heavy fluid seeps through the bottom of the barrier and begins to form a gravity current front, while at the same time a weak hydraulic jump propagates upstream from the barrier. The splash associated with the initial encounter is much weaker, as expected, than in the case of the solid barrier. Eventually, a steady state is

Figure 11.20 The laboratory interaction of a gravity current with a porous obstacle. This consists of four rows of ten rods, 10 mm in diameter and 10 cm high.

achieved, with a rapid drop in the interface level through the barrier with a gravity current front propagating at a lower speed downstream from the barrier.

11.6.3 Flow under a barrier

Figure 11.21 shows the form of the head of a laboratory gravity current which has just passed beneath a sharp barrier across the upper half of a two-dimensional channel. This current was generated by a lock exchange and as it approached the barrier the dense fluid occupied about half the depth of the tank. The shadowgraph shows that the depth of the current was reduced to about half the height of the gap beneath the barrier.

One feature which is clearly shown in this shadowgraph is the turbulence formed above the dense current as it approaches the barrier. The reverse flow through the restricted space breaks up into a turbulent zone filling most of the space up to the surface.

A rather similar environmental problem which can be examined in the laboratory is that of a power station effluent in which a gravity current of warm water on the surface of the sea passes above a submarine ridge. To avoid problems of a contaminated free surface and uncontrolled heat losses, the laboratory experiment shown in figure 11.22 has been carried out, in which the flow and obstacle have been inverted. A dense salt solution moved beneath a smooth obstacle, shaped as shown, and a series of photographs served to record the progress of the front and to show its changes in form.

As the gravity current passed beneath the obstacle it became reduced to half its original height, and measurements showed that its speed was reduced by the factor of $(1/2)^{\frac{1}{2}}$.

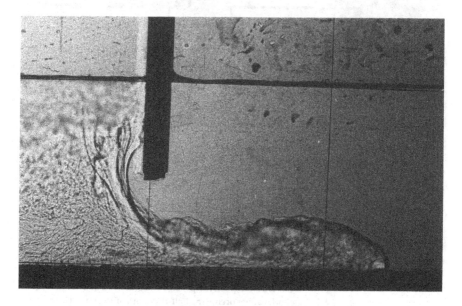

Figure 11.21 The form of the head of a gravity current which has just passed beneath a sharp barrier.

11.6.4 Gravity current in steadily decreasing depth

Experiments on gravity currents flowing into fluid of steadily decreasing depth are best carried out with dense currents running along the floor, beneath a rigid sloping lid. As in the previous section, surface contamination effects can be avoided, and it is possible to apply a small correction to the results for flows in the ocean at a free surface, above a rising sea bed.

Each gravity current was released from a lock at the end of the tank and

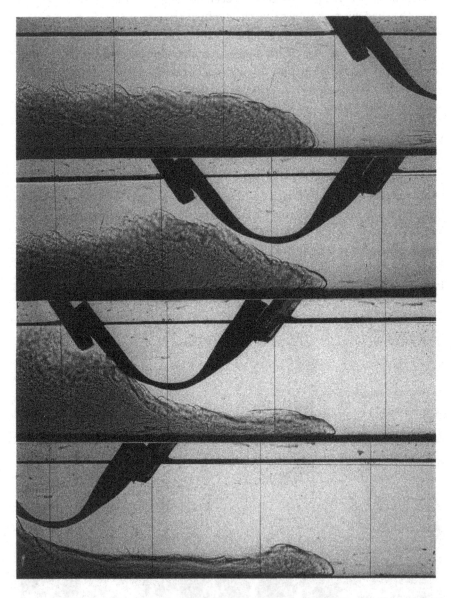

Figure 11.22 Stages in the passage of a gravity current beneath a smooth barrier occupying half the depth of the tank. The lines are 10 cm apart, and the times since release are 8.4.11.5.14.4 and 17 seconds.

allowed to flow beneath a rigid flat ceiling sloping down in the direction of the current. The angle of slope varied from 1 degree to 21 degrees.

When the depth of the gravity current head became as large as half the local depth below the lid, the breaking head-waves disappeared and the current interface downstream of the head became roughly parallel to the ceiling. The photographs in figure 11.23 show two stages of a gravity current beneath a lid with a slope of about 5 degrees. The depth of the dense fluid at the front remains about half the local depth, so that the depth of the parallel-sided space above the liquid continues to decrease as the front advances. The second photograph shows almost complete absence of mixing at the interface.

Measurements were made of the speed U of the front for different values of the horizontal distance x from the intersection of the ceiling and the floor, for different ceiling angles. Figure 11.24 shows values of U^2 plotted against $g'x \tan \theta$ and it can be seen that the points lie close to a straight line through the origin, giving $U^2 = 0.21\, g'x \tan \theta$, or $U = 0.45(g'x \tan \theta)^{\frac{1}{2}}$.

As the form of the head under the sloping lid resembles the non-mixing head seen at low Reynolds number gravity currents, tests were made to see whether the non-mixing head was a low Reynolds number effect. In one such test

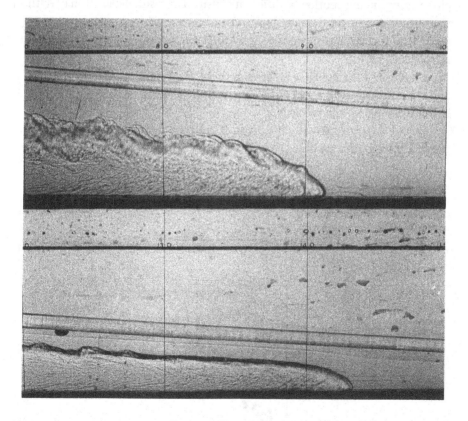

Figure 11.23 Two stages in the flow of a gravity current beneath a sloping lid. The lines are 10 cm apart, and the times since release are 17.4 and 25 seconds.

the sloping lid was made parallel to the floor before the end of the tank; the usual head-structure with breaking waves appeared beneath the level lid. The least slope which would suppress head-mixing was found to be about 1/80, and all the experimental values of the Froude number $U/(g'x \tan \theta)$, for slope greater than 1/80, were plotted against the Reynolds number $Re = U x \tan \theta/\nu$. No trend with Reynolds number could be detected in the range from $Re = 50$ to 5000 examined.

If we write $H = x \tan \theta$, the total height of the space immediately above the front of the current, then it appears that $U = 0.45(g'H)^{\frac{1}{2}}$ and the current at any time adjusts almost instantaneously to the same value as it would have underneath a level lid of the same height, since $U = 0.45(g'H)^{\frac{1}{2}}$ is the usually accepted value for the speed in the initial constant speed regime in a lock-exchange flow of total depth H.

11.7 Purging of fluid in a pipe

An important industrial problem is the displacement of a fluid in a horizontal pipe by introducing another of different density. The result depends on both the rate of introduction of the new fluid and the buoyancy forces acting on the two fluids. Figure 11.25 shows three stages in which the rate of input of the new fluid is increased.

At low input rates ('Flow I') the buoyancy force dominates and the fluid moves along the pipe as a gravity current with some turbulent mixing at the head – a dissipative gravity current.

As the flow rate increases the depth of the fluid increases until it reaches the maximum depth possible ('Flow II'), which is determined by the shape of the

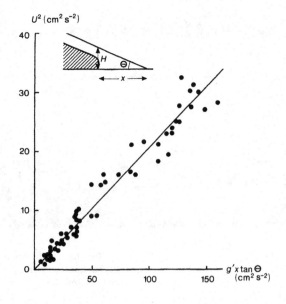

Figure 11.24 The square of the speed of the front, U, plotted against distance, x, for gravity currents travelling beneath a sloping lid.

Figure 11.25 Stages in the purging of a fluid through a duct. Flow I. Simple dissipative gravity current. Flow II. Loss-free gravity current followed by a hydraulic jump. Flow III. Complete purging: following the wedge, or expansion wave, the flow expands to the bottom of the channel.

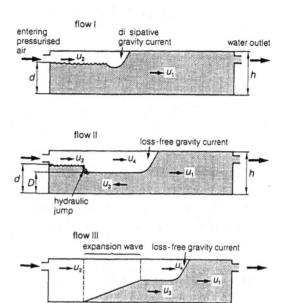

duct, being 0.5 in a rectangular channel, but 0.62 in a circular pipe. This marks the largest flow rate possible by a gravity current which is then non-turbulent and loss-free.

For higher input rates ('Flow III') the input fluid cannot be carried as a gravity current and the front is followed by a trailing edge as the input fluid then displaces all the resident fluid.

Bibliography

Barr, D.I.H. (1967). Densimetric exchange flows in rectangular channels. *Houille Blanche*, **22**, 619–31.

Benjamin, T.B. (1968). Gravity currents and related phenomena. *J. Fluid Mech.*, **31**, 209–43.

Britter, R.E. & Simpson, J.E. (1978). Experiments on the dynamics of a gravity current head. *J. Fluid Mech.*, **88**, 223–40.

Greenspan, H.P. & Young, R.E. (1978). Flow over a containment dyke. *J. Fluid Mech.*, **87**, 179–92.

Keulegan, G.H. (1957). An experimental study of the motion of saline water from locks into fresh water channels. *U.S. Natl. Bur. Stand. Rep.* 5168.

Lawson, T.J. (1971). Haboob structure at Khartoum. *Weather*, **26**, 110–12.

Riddell, J.C. (1970). Arrested saline wedge. *Houille Blanche*, **4**, 317–30.

Rottman, J.W., Simpson, J.E., Hunt, J.S.R. & Britter, R.E. (1985). Unsteady gravity currents over obstacles. *J. Hazardous Mat.*, **11**, 325–40.

Schmidt, W. (1911). Zur Mechanik der Boen. *Z. Meteorol.*, **28**, 355–62.

Simpson, J.E. (1969). A comparison between laboratory and atmospheric density currents. *Q. J. R. Meteorol. Soc.*, **95**, 758–65.

Simpson, J.E. (1972). Effects of the lower boundary on the head of a gravity current. *J. Fluid Mech.*, **53**, 759–68.

Simpson, J.E. & Britter, R.E. (1979). The dynamics of the head of a gravity current advancing over a horizontal surface. *J. Fluid Mech.*, **94**, 477–95.

Simpson, J.E. & Britter, R.E. (1980). A laboratory model of an atmospheric mesofront. *Q. J. R. Meteorol. Soc.*, **106**, 485–500.

Thorpe, S.A. (1973). Experiments on instability and turbulence in a stratified shear flow. *J. Fluid Mech.*, **61**, 731–51.

Von Karman, T. (1940). The engineer grapples with non-linear problems. *Bull. Am. Math. Soc.*, **46**, 615.

Winant, C.D. & Bratkovich, A. (1977). Structure and mixing within the frontal region of a density current. *6th Aust. Hydraul. & Fluid Mech. Conf., Adelaide*, pp. 9–12.

12   Spread of dense fluid

The previous chapter considered in detail the advance of the front of a gravity current when its depth and density difference have already been established. This front plays an important part in the dynamics of a gravity current, but to deal with the flow as a whole, one needs to know how the fluid is being supplied and the form of the surroundings into which it is allowed to spread.

12.1   Lock exchange flows

Some of the earliest measurements of gravity currents were made in navigation canals near the coast where exchanges of fresh and salt water always took place whenever a lock gate was opened.

To set up an experimental system, a channel is temporarily divided into two sections by a thin vertical barrier. Fresh water is run into one segment and salt water into the other, and the levels are made equal. As soon as the barrier is raised, the dense fluid starts to collapse and counter currents begin to flow in opposite directions. These consist of a gravity current of less-dense fluid moving along the surface and a dense current flowing beneath it in the other direction.

Many experiments of this nature have been carried out both in canals (O'Brien & Cherno, 1934) and on a smaller scale in the laboratory (Yih, 1980). Lock exchange flows in a channel of uniform rectangular cross-section will be considered first. The most notable feature of these flows is the uniformity of the front velocity. The front continues to advance at a constant speed, provided that the two parts of the channel are both very long and that the flow does not reach the range of viscous control (see Chapter 15).

Several authors have given the same analysis, in which the extension of a rectangular block as in figure 12.1a is assumed as a first approximation to lock exchange flow. By equating decrease of potential energy with increase of kinetic energy, the following result is obtained:

$$U_0/(g'H)^{\frac{1}{2}} = 0.5$$

This forecast is very close to the value of 0.46 for the velocity found in an extended series of experiments in very large tanks (Barr, 1967).

The photograph in figure 12.1b is of a developing lock exchange flow in the laboratory. This shows some features which are different from the simple theoretical model. Each flow has a head slightly deeper than the following flow and the flow has a slight slope behind it.

The front along the free surface advances slightly faster than that along the lower boundary and a value of 0.59 has been found for the non-dimensional velocity of the overflow in these experiments. In section 11.6 the difficulties in obtaining consistent results in such experiments with surface flows have already been noted, but the experiments here were carried out in very large tanks, where unaccountable surface effects should be small.

Although the experimental results agree fairly well with the theoretical forecast for the overall advance of the front, in fact, the assumption of block flow makes no allowance for the maximum internal velocities. These have been found experimentally to be of the order of 1.2 times the front velocity; the relationship with the mixing at the leading edge has been investigated by later experimenters (Barr & Hassan, 1963; Simpson & Britter, 1979).

12.1.1 Trapezoidal, circular and other cross-sections

Some experiments carried out in trapezoidal flumes can be seen to fit quite well into the theoretical pattern.

In the case of the overflow front in a triangular section with a free surface, the initial velocity is related to the maximum depth by

$$U_0/(g'H)^{\frac{1}{2}} = 0.40$$

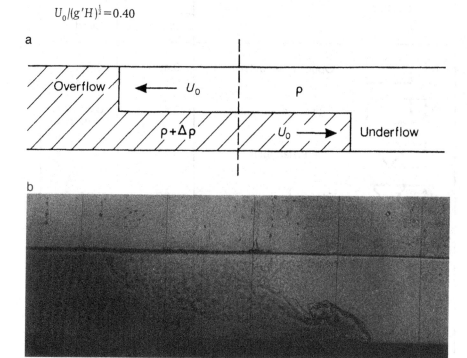

Figure 12.1 Developing lock exchange flow. (a) Inviscid, non-mixing theoretical model. (b) Shadowgraph of flow in a laboratory experiment.

Between rectangular and triangular sections there is a reversal of the comparative pattern of underflow and overflow. Exchange flows in cylindrial tubes are clearly also of importance and figure 12.2 gives values of the non-dimensional initial velocities in channels of rectangular, triangular and circular sections.

12.2 Release of a fixed quantity of fluid in a rectangular channel

The virtue of lock exchange experiments is their relative simplicity, making it possible to perform large numbers of experimental runs to illustrate the whole range of gravity currents, including those in which viscous effects have become dominant. In many experiments the volume of one of the fluids is made much smaller than that of the other, so that the procedure approximates to the release of a finite volume of fluid into another fluid of infinite volume.

It has been shown experimentally that a gravity current produced by instantaneous release passes through two distinct phases. (A third phase may be reached if viscous effects become dominant.) After the rapid initial collapse

Section		Non-dimensional initial velocity	
		Underflow	Overflow
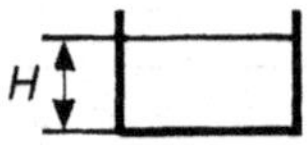 H Open rectangular	Open rectangular	0.465	0.59
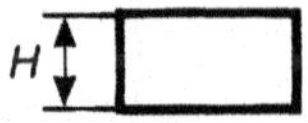 H Closed rectangular	Closed rectangular	0.44 approx	0.44 approx
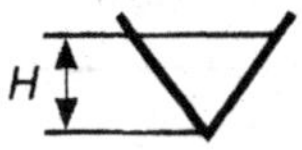 H Open triangular	Open triangular	0.67	0.41
 H Closed triangular	Closed triangular	0.65	0.34
 H Circular pipe (full)	Circular pipe (full)	0.495	0.495
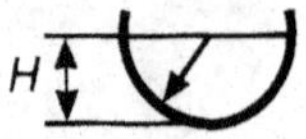 H Half depth circular	Half depth circular	0.51	0.44
 H Quarter depth circular	Quarter depth circular	0.48	0.36

Figure 12.2 Experimental values of the initial velocity of gravity current fronts in channels of different cross-section. The velocity is given as $U/(g'H)^{\frac{1}{2}}$. (Courtesy of D.A.H. Barr.)

when the gate is removed there is an adjustment phase in which the front advances at constant speed (the first phase). In this phase the initial conditions are important. This merges into an eventual phase in which the front speed decreases as $t^{-\frac{1}{2}}$ (the second phase) (where t is the time measured from release). It may be more convenient to express this relationship in the form 'the distance, x, varies as $t^{\frac{1}{2}}$'. The transition from the first to the second phase is observed to be rather abrupt.

12.2.1 First phase (constant speed)

It has been shown that the transition from the first to the second phase occurs when a disturbance generated at the end wall (or plane of symmetry) overtakes the front.

The nature of the travelling disturbance differs in the following two cases: (a) those in which the segment of dense fluid has the same depth as the rest of the fluid; and (b) where the dense fluid released is shallower than the total depth of the fluid, as in the 'dam-break analogy' experiments.

In the former case the depth, h_0, of the dense fluid is equal to that of the rest of the fluid, H, in the channel. As soon as the gate is removed, the fluid from behind the gate forms a gravity current with structure as described above. The intense mixing between the two fluids is confined to a region just behind the leading edge of the current, the mixed fluid being left behind the head and above the following current. At the same time the displaced upper fluid forms a gravity current that propagates towards the end wall (see figure 12.3). When the backflowing current meets the wall, a hydraulic drop is generated (Rottman & Simpson, 1983). This may be pictured as an 'inverted bore' consisting of the lighter fluid advancing through the upper part of the dense fluid beneath it. This disturbance propagates away from the wall, and eventually overtakes the front. Some measurements of the position of the front of a volume of salt solution released from a lock in which $h_0/H = 1$ are shown in figure 12.4. In these experiments the front was observed to be about ten lock lengths from the end wall when it was overtaken by the bore. After this stage is reached the speed of the front is no longer constant and decreases with $t^{-\frac{1}{3}}$ (or x varies with $t^{\frac{2}{3}}$).

The behaviour of a 'dam-break analogy' gravity current in deep water is different. Figure 12.5 illustrates the difference from the lock exchange in which a bore is formed; this is shown in figure 12.5a. Figure 12.5b shows the stages in the collapse of a volume of heavy fluid released into deep water, in which h_0/H approaches zero. As soon as the gate is removed, the dense fluid forms a gravity current moving away from the end wall at constant speed, as in the previous case. At the same time a long wave of depression propagates along the fluid interface towards the left. This wave is reflected by the end wall ((iii) in figure 12.5b) and propagates away from the wall with speed slightly greater than the speed of the front ((iii) in figure 12.5b), eventually overtaking the front ((iv) in figure 12.5b). Thereafter, the front speed, which had been constant up to this

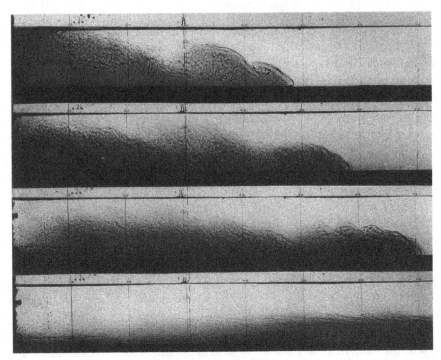

Figure 12.3 Shadowgraph of a volume of salt water collapsing into fresh water, at 5, 7, 9 and 17 seconds after release. The dotted vertical line shows the position of the lock gate. The vertical lines are 10 cm apart and the end wall can just be seen on the left.

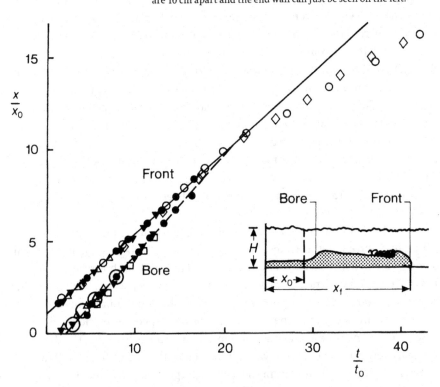

Figure 12.4 The position of a gravity current front which is being overtaken by a bore reflected from the end wall of a tank, from a large series of experiments.

point, decreases as $t^{-\frac{1}{3}}$ until viscous effects become more important than inertial effects, causing the front speed to decrease more rapidly.

The shallower the upper layer is relative to the dense layer, the greater must be the speed of the return flow there. Theory indicates that this trend results in the formation of an interfacial hydraulic drop for $h_0/H > \frac{1}{2}$, but experiments give no indication of a hydraulic drop until $h_0/H > 0.7$. The values of the constant speed during the first stage for other values of h_0/H than 1 have been shown to decrease almost linearly from 0.7 when $h_0/H = 0$ to about 0.5, as already mentioned, for $h_0/H = 1$.

12.2.2 Second stage (self-similar flow)

During the stage when the front speed has begun to decrease, the gravity current is well described as collapsing through a series of equal-area rectangles, the so-called box model, in which the current depth is roughly uniform along the length of the current, but steadily decreases with time.

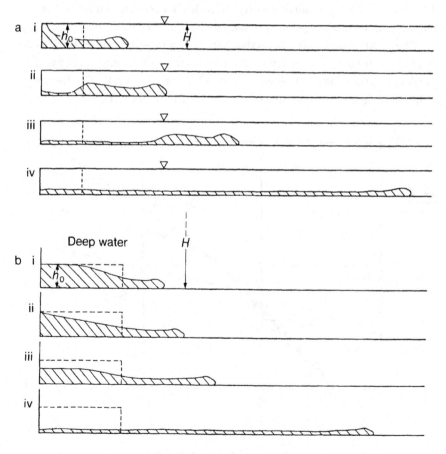

Figure 12.5 The collapse of a volume of dense fluid at four stages after release. The dense fluid is the same depth as the fresh water in (a) and is much less deep than the water in the channel in (b). The dashed lines indicate the boundary of the fluid before collapse.

On the grounds that the length of the current greatly exceeds its vertical thickness, an analysis has been based on the depth-averaged shallow-water equations (Baines *et al.*, 1985).

Retaining only the buoyancy and inertial terms, the equations have been solved and the coefficients evaluated from various experiments, including some on the spreading of oil over water. Distance, X,

$$X = 1.6(g'q)^{\frac{1}{3}}t^{\frac{2}{3}}$$

was obtained for the two-dimensional case. This result has already been mentioned above as following the initial constant speed regime, and figure 12.6 gives some experimental results (Rottman & Simpson, 1983) showing the position of the front with time. In this log–log plot the distance $(x_t - x_0)$ is non-dimensionalised by the initial lock-length, x_0, and the time t by $t_0 = (x_0/g'h)^{\frac{1}{2}}$. It can be seen that the points during the initial adjustment phase are on the line with slope 1, showing uniform velocity. In the inviscid self-similar phase the points are on the line with slope $\frac{2}{3}$, showing distance varying with $t^{\frac{2}{3}}$.

The points representing flows that have begun to be dominated by viscosity can easily be distinguished when they reach this stage in the graph. They then slow down further and can be seen on lines with slope $\frac{1}{5}$, a result to be discussed further in Chapter 15 in the context of viscosity-dominated currents.

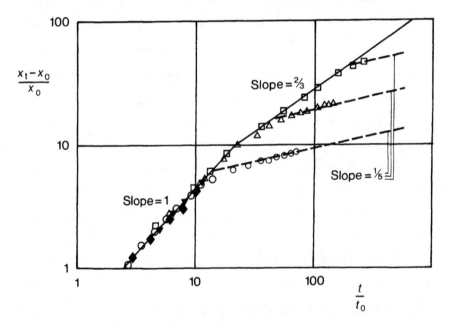

Figure 12.6 Log–log plot of the change in the position of four different gravity fronts with time. The change of gradient from 1 to $\frac{2}{3}$ occurs when the distance is ten lock-lengths. The three currents which attain at successive times the viscous regime change to a gradient of $\frac{1}{5}$ are shown.

12.2.3 Volume of air collapsing into water

Experiments in which a fixed volume of air is released into a closed horizontal tank full of water are illustrated in figure 12.7. Such experiments show clearly the stages in the development of a 'gravity current' of air, after release from behind a gate, as it progresses above the water.

In phase A the front is moving at a constant speed and occupies almost exactly half the depth of the tank. There is no turbulence at the head in this constant-speed phase with no energy loss, which was described theoretically in Chapter 11 and illustrated in figure 11.4.

In the second photograph, B, an advancing hydraulic jump, or bore, is seen on the left-hand side. This turbulent step in height of the water is steadily catching up with the front.

In the third view, C, the bore has reached the head and a roughly level current of air exists, showing some turbulence at the lower interface with the water. The speed in this phase is no longer constant.

The results of four experiments (not shown here) plot distance against time. The distance, X, is non-dimensionalised by X_0, the lock-length and the time

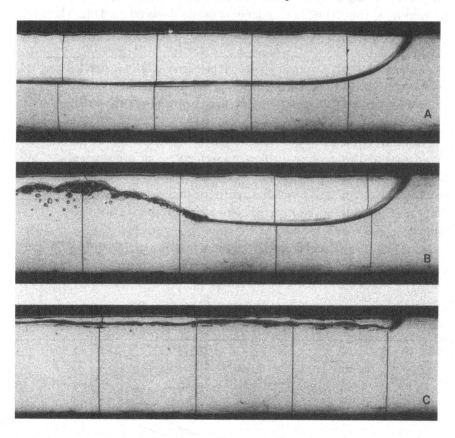

Figure 12.7 Three stages of the form of an advancing volume of air which has been released from one end of a channel of water. The depth of the tank is 10 cm.

by t_0, which is $(H/g')^{\frac{1}{2}}$. The results from the experiments all fall on the same lines. The bore is seen to be derived from the reflection from the end wall of the tank, and the end of the constant speed regime starts when the bore has reached the front of the air flow.

There is also a third phase which may be reached if the flow continues far enough. In this phase the cavity motion becomes erratic due to the dominance of surface tension forces (Baines et al., 1985). The surface tension forces become strong enough to bring the motion to rest, i.e., the cavity approaches the limit of a static two-dimensional bubble. The motion becomes erratic because the randomly distributed contaminants on the plexiglass surface affect the contact angle.

12.3 Radial collapse of a fixed quantity of fluid

A fixed quantity of dense fluid released into another fluid of different density, with no restraining barriers, spreads out in all directions. In early experiments by Penny & Thornhill (1952) a cylinder of dense fluid was released into a larger tank. After a rapid initial collapse a roughly axisymmetrical spread of the fluid was observed.

After an initial adjustment phase, but before a viscosity-controlled stage is reached, a cylindrical 'box model' is appropriate and it has been shown (Hoult, 1972; Fay, 1969) that the distance, X, travelled by the front varies with the time, t, to the power of $\frac{1}{2}$.

Later experiments on axisymmetrical flows have been made using tanks as shown in figure 12.8 in which the flow in a sector of about 10 degrees of a cylinder is observed. This uses much less volume of fluid for any given distance travelled, and has the advantage that in a transparent sector tank the profile of the head of the front can be observed more accurately than in a complete cylinder. It is also possible to take shadowgraph photographs.

The line of the front becomes nearly straight as a radial flow develops, and, as might be expected, the head then behaves very similarly to that observed in a

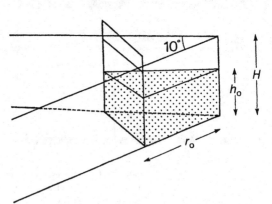

Figure 12.8 Tank in the form of a sector for the release of a fixed volume of dense fluid to form an axisymmetrical gravity current.

flow between parallel walls. One difference is that the distance from the start is proportional to $t^{\frac{1}{2}}$ instead of $t^{\frac{2}{3}}$.

12.3.1 Vortex ring at start of collapse

In the initial phase of adjustment, however, there are important differences between two-dimensional and axisymmetric flows. Figure 12.9 shows sequential shadowgraphs of the initial collapse of salt water in a sector-shaped tank. In figure 12.9a the initial fractional depth h_0/H is $\frac{1}{4}$, and in figure 12.9b $h_0=H$. In all these flows, and indeed also in the two-dimensional releases, the front forms very soon after release (in a few tenths of a second) and the most intense mixing of the two fluids occurs near the front. The most striking difference between the axisymmetric and two-dimensional flows is the intensity of rotational motion of the internal fluid during the early stages. In the time taken by the axisymmetric front to travel a distance equal to about one lock-length, the majority of the fluid in the current becomes concentrated at the front (for small h_0/H) or in multiple fronts (for h_0/H near to 1), leaving only a thin layer of heavy fluid near the ground. In contrast, the two-dimensional flows are more uniform in depth shortly after release. In addition, the internal rotational flow and associated mixing is more intense; indeed, the mixing appears to occur all the way down to the ground behind the front (or fronts) in these flows.

The formation of this vortex ring during the early stages of spreading out of a dense fluid has important applications to aircraft safety and was dealt with in more detail in Chapter 6. The details of this important vortex development are shown in figure 12.10. The pictures show that by stage c the vortex has extended down close to the lower boundary.

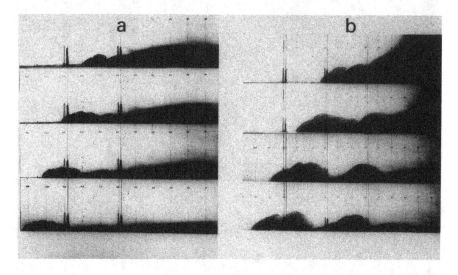

Figure 12.9 Shadowgraphs of a volume of salt water collapsing into fresh water in a sector-shaped tank. The total depth is 40 cm and reduced gravity, g', is 47 cm s^{-2}. In (a) $h_0/H=0.25$ and in (b) h_0/H is 1.

We know that Kelvin–Helmholtz vortices form as the leading edge of
a gravity current develops, and because the fluid volume in a vortex is
approximately conserved, its cross-sectional area must decrease as it stretches.
Conservation of angular momentum about the centre line of the vortex then
implies that its intensity increases. This intensification is largest during the rapid
expansion near the source and produces a leading-edge vortex which occupies
almost the full depth of the dense fluid.

To show that the vorticity already being produced at the leading edge of
an advancing gravity current is sufficient to produce this intense roll-up of the
leading edge when it is stretched, a simple experiment has been carried out
(Rottman & Simpson, 1984). In this experiment a saline gravity current
(produced by lock exchange) flowing between parallel walls was suddenly
allowed to double its width. Figure 12.11 shows the simple apparatus. The
behaviour of the leading edge of the resultant gravity current is shown in figure

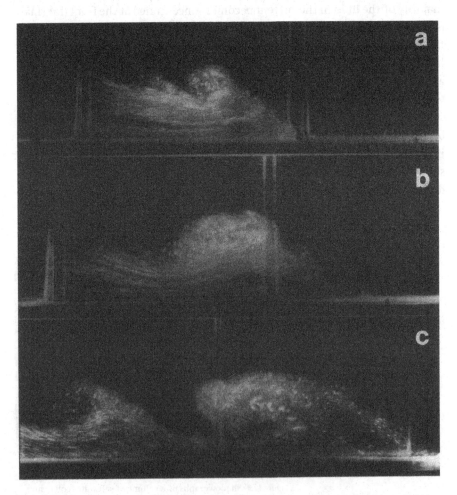

Figure 12.10 The collapse and spread of a volume of dense fluid in
a sector-tank, made visible by fine aluminium particles. Time
intervals are two seconds and $g' = 12$ cm s^{-2}.

12.12. The leading-edge vortex rolled up as the width of the front increased and reached almost down to the ground in figure 12.12b.

12.4 Constant flux in a parallel channel

Consider a gravity current in a parallel-sided channel which is being supplied by a steady input of dense fluid. This input contains a 'buoyancy flux', where the

Figure 12.11 Tank to test the effect of widening on the form of the head of a gravity current that was originally in a channel with narrower parallel sides.

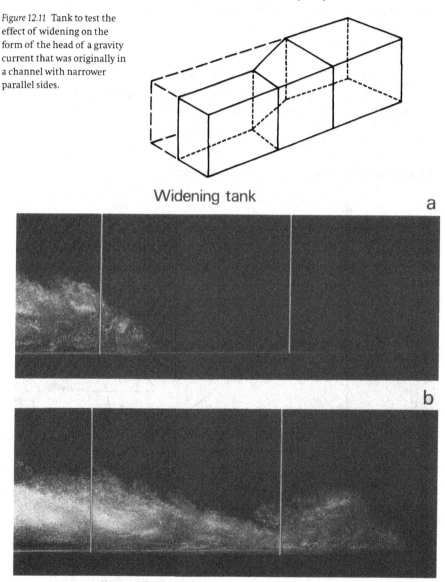

Widening tank

a

b

Figure 12.12 The effect of widening the channel on the form of a gravity current head. The width increases from 10 cm to 20 cm between the two white lines, which are 30 cm apart. The time interval is 5.7 s, and g' is 14 cm s^{-2}.

buoyancy may be either positive or negative, and the gravity current
consequently either at a free surface or along the ground. An example is the
warm outflow from a power station where both the advance of the initial
leading edge and the mixing during later stages are of interest.

These so-called 'starting plumes' are different from the lock exchange flows.
In lock exchanges there is no total flow across any cross-section of the container
of the fluid, so that a gravity current of large fractional depth must have a faster
return flow above it than will a starting plume of the same depth. For inviscid,
non-mixing, starting flows the Froude number equals

$$[(2-\varphi)/(1-\varphi^2)]^{\frac{1}{2}}$$

where ϕ is the fractional depth h/H of the current.

As shown in the upper curve of figure 12.13 this expression varies much
less with than the corresponding one

$$[(2-\varphi)(1-\varphi)/(1+\varphi)]^{\frac{1}{2}}$$

for lock exchange flows, shown in the lower curve.

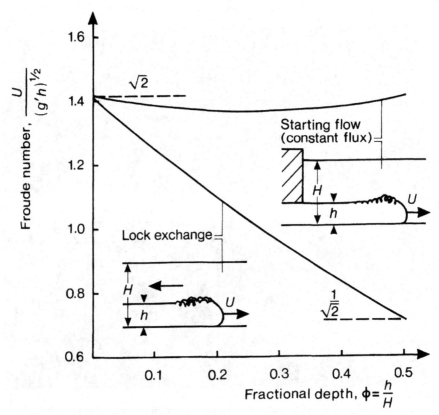

Figure 12.13 Comparison between the speed of the front of a lock
exchange flow and that of a constant flux, at the same fractional
depth.

It is possible to set up an experimental flow rate in which the layer stays uniform for some distance along the flume, but for increasing distances the velocity decreases. It has been shown that for an inlet Froude number greater than 1 there exists an entraining hydraulic jump at the interface between the fluids, as shown in figure 12.14, and it seems that the rate of entrainment here and the conditions downstream are controlled by the gravity current head.

This hydraulic jump, or 'density jump' as it has been called (Wilkinson & Wood, 1972), is able to entrain the varying amounts of ambient fluid necessary to satisfy a range of downstream conditions. However, along a horizontal floor with a gravity current head no equilibrium state exists (Koh, 1971) and the velocity of the front continues to fall, eventually causing the density jump to flood.

12.5 Constant flux, radial flow

Analogous results to the two-dimensional flow have been obtained for axisymmetric flows. Such flows are of interest in describing several geophysical events such as the spread through a narrow gap of river water at the sea surface.

The spread of such a plume has been analysed and described by Chen & List (1976) and by Britter (1979), in separate regimes, first dominated by an inertia–buoyancy force balance and then by a friction–buoyancy balance. The rate of spread for the first regime was shown to be given by

$$R(t)=k_1(Qg')^{\frac{1}{3}}t^{\frac{1}{2}}$$

where $k_1=0.75$ and Q is the mass flow per unit width. Experiments (Britter, 1979) give a value for k_1 of 0.84.

A gravity current which is spreading out after passing through a narrow slot often shows periodic oscillations in the form of the head. These oscillations

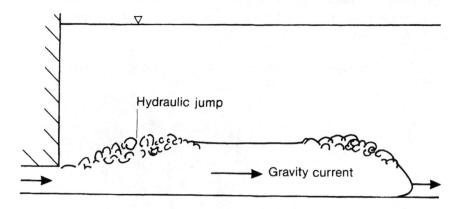

Hydraulic jump

Gravity current

Figure 12.14 Gravity flow with inlet Froude number greater than one, showing the entraining hydraulic jump.

have been seen in environmental flows and examples have been observed in laboratory experiments. Figure 12.15 shows an example of oscillations seen in plan form as a gravity current which is entering through a small gap spreads out over the ground (Linden & Simpson, 1985).

12.6 Front moving along a free surface

Plumes of fresh water flowing above salt sea water are examples of gravity currents moving along a surface free to the atmosphere. Modelling these flows in the laboratory needs care with the selection of scale.

Small-scale laboratory buoyant gravity currents, sometimes called 'overflows' as opposed to 'underflows' along the ground, can be strongly affected by surface forces. These are caused by surface tension and by films resulting from surface-active materials in the water. These effects may be large and it is difficult to reduce them, but experiments by Thomas & Simpson (1983) have shown that surface-tension effects are small provided that $\Sigma = \Delta\sigma/g'h$ ($\Delta\sigma$ is the surface-tension difference) is smaller than 0.01, and that surface film effects are not important when the current depth is large compared with the boundary-layer depth.

12.7 Gravity currents on slopes

Not all gravity currents in the environment flow along horizontal surfaces. For example, the sloping surface of a mountain plays a vital part in the formation and development of a snow avalanche. A slope is also needed for the formation of the turbidity currents which form in the ocean at the edge of the continental shelf.

From direct measurements of the behaviour of buoyant methane gas along the roof of a passage of a mine (Georgeson, 1942) and from laboratory experiments (Middleton, 1966; Hopfinger & Tochon-Danguy, 1977), it is clear that the motion of a gravity current down a slope is appreciably different from that along a horizontal surface.

Figure 12.15 View from above of a constant-flux gravity current spreading out on the ground, giving evidence of oscillations.

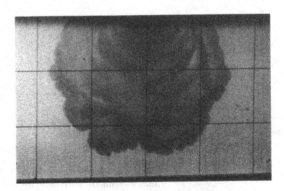

For a given rate of flow of dense material on slopes from a few degrees to 90 degrees there is a nearly constant front velocity. This velocity is found to be about 60% of the mean velocity of the following flow, or, in other words, about 40% of the flow arriving at the front is mixed into the head.

Figure 12.16 shows three different forms of the head of a gravity current, first on a horizontal surface and then down slopes of 5 degrees and 20 degrees, respectively. The head volume increases, both by direct entrainment and also by addition from the following flow. There is little variation of front speed with slope because, although the gravitational forces increase with slope, so does the entrainment, both into the head and into the flow behind it.

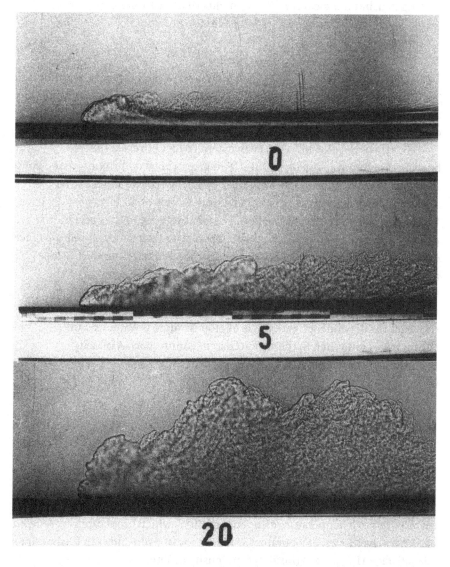

Figure 12.16 Gravity currents flowing down slopes of 0 degrees, 5 degrees and 20 degrees. The entrainment increases with slope, both into the head and into the flow behind it.

The amount of direct entrainment increases with slope, leading to $\frac{1}{10}$ of the head-growth at 10 degrees and about $\frac{2}{3}$ at 90 degrees. With a steady buoyancy flux, Q, experiments by Britter & Linden (1980) showed a constant head velocity for slopes greater than 5 degrees, and from 5 to 90 degrees the front velocity was given by

$$U/(g'Q)=1.5\pm0.2$$

The transition zone between steady currents down a slope and time-dependent currents along a horizontal surface has not yet been thoroughly investigated, but it is clear from experiments that the front behaviour is extremely sensitive to slope as it approaches the horizontal.

12.7.1 Motion of an 'inclined thermal'

The release of a finite volume of dense fluid down a slope, and the resulting 'inclined thermal' of 'fixed-volume cloud' has practical applications. For example, the study of this problem is as relevant to avalanches as the case of constant flow.

The theory of thermals has been extended (Beghin *et al.*, 1981) to buoyant cloud convection on inclines. This gives a good description of the flow in the range from 5 degrees to 90 degrees. The shape of the cloud can be approximated by an ellipse after a characteristic time of acceleration, followed by a deceleration of the flow. The front velocity in the decelerating phase, normalised by the square root of the released buoyancy and the distance of the leading edge from the virtual origin, is plotted as a function of slope angle in figure 12.17. Its behaviour can be compared here with that of the front of the steady current, discussed above. The experimental scatter in results is large, due to the variability in structure of the front, but it can be seen that the dependence with angle of slope is more pronounced in the thermal cloud than in the steady current counterpart. A velocity maximum in the 'thermal' is again observed at slope angles between 10 and 40 degrees.

12.8 Gravity currents of high density ratio

Most data on the propagation of gravity current fronts have dealt in the past with low density ratios. In the environment the limit of density excess in saline flows in the ocean is about 3%; in the atmosphere this density difference corresponds to a temperature difference of $10\,^{\circ}$C, which is not often exceeded in boundary layer flows. In laboratory work the use of saline flows in fresh water has satisfactorily covered this range of density differences.

Much higher density ratios are found in the accidental release of dense industrial gases, and in the spread of hot fire gases along ceilings. In the 1990s disagreements were found during the measurements of this type of flow involving gravity currents with density ratios as high as 5 or even 10.

In dealing with these high density ratios some of the simplifications which have been taken for granted are no longer justified. One of these often quoted is the Boussinesq approximation; as applied to gravity currents this consists of neglecting variations in density as far as they affect inertia, but retaining them in the buoyancy terms, where they occur in combinations of diluted gravity, $g' = g\Delta\rho/\rho$.

In currents of high density ratio we can no longer make the Boussinesq approximation and must treat separately the 'heavy' and 'light' flows in a lock exchange.

If the densities of the light and heavy fluids are ρ_1 and ρ_2 respectively:

for the 'Light', we write $g' = g(\rho_2 - \rho_1)/\rho_2$
and for the 'Heavy', $\qquad g' = g(\rho_2 - \rho_1)/\rho_1$

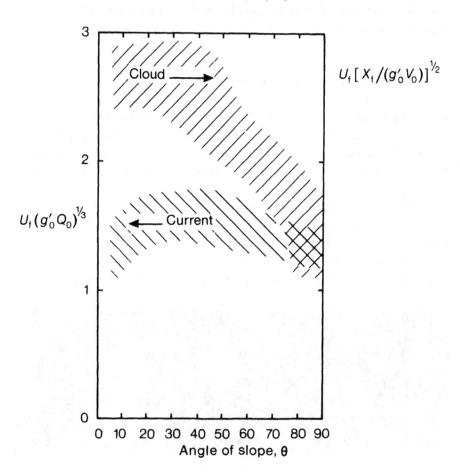

Figure 12.17 Comparison of the speed down a slope of a continuous gravity current and a 'cloud' or 'thermal'.

These two flows have different Froude numbers $U/(g'h)^{\frac{1}{2}}$; for large density ratios the nature of the fronts in the top and bottom flows may be very different, as shown in figure 12.18. This shows an exchange flow between helium and air, which have a density ratio of 7.2. The differences in the form of the two fronts can be seen: to the right the upper front of helium is quite smooth, while the denser air front has the usual raised head and following billows.

The graph in figure 12.19 shows the result of a series of similar exchange flows between gases of different density ratio, made at a fractional depth, $\phi=\frac{1}{2}$. The density ratios are given in the form

$$\rho^* = [(\rho_2 - \rho_1)/(\rho_2 + \rho_1)]^{\frac{1}{2}}$$

and the Froude number $\mathrm{Fr} = U_f/(g'h)^{\frac{1}{2}}$ is plotted for ρ^* from 0 to 1, showing the different values for the 'light' and the 'heavy' gases.

For the extreme value of $\rho^* = 1$, Froude numbers of $1/\sqrt{2}$ and $2\sqrt{2}$ have been plotted for the 'light' and 'heavy' flows respectively. The light 'Benjamin flow' was shown in Chapter 11, in figure 11.4 and the heavy 'dam-break' front appears in Henderson (1966).

12.9 Gravity currents over porous media

Sometimes the ground over which a gravity current moves may not be a solid impervious surface and has to be treated as a porous medium. An example might be the spill of a reservoir of LNG (Liquid Natural Gas) when most of the container is surrounded by porous ground; this will affect the nature of the spread of the dense gas.

Laboratory experiments have been carried out (Lionet & Quoy, pers. commun.) in which a gravity current of salt solution travels through fresh water along a porous floor. Figure 12.20 shows the apparatus used for such an experiment and one of the flows in action. To start an experiment, the gate containing the salt water is partly raised; this releases a constant supply of dense

Figure 12.18 Exchange flow between air and helium. The density ratio is 7.2. Photo by J. Muller.

fluid which moves above a porous barrier consisting of two metallic grids with small holes, one above the other.

A simple theory, employing a 'box model', gives the velocity of the front as a linear decreasing function of time, confirmed by the experimental results. The other important result is that, since the front of the gravity current must come to rest when all the mass has disappeared, a maximum distance will be attained. This was found to vary with $1/(g')^{\frac{1}{2}}$.

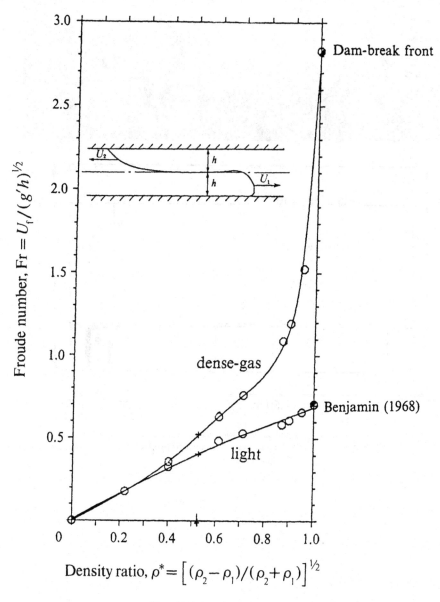

Figure 12.19 Froude number of light- and dense-gas density ratios. Fractional depth $\phi=\frac{1}{2}$ (From Gröbelbauer *et al.*, 1993; Courtesy of T. Fannelop.)

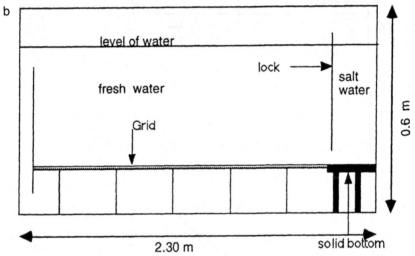

Figure 12.20 Laboratory gravity current of salt water in a tank above a porous floor, consisting of two grids with holes. (Courtesy of J. Lionet & O. Quoy.)

Bibliography

Baines, W.D., Rottman, J.W. & Simpson, J.E. (1985). The motion of constant volume air cavities in long horizontal tubes. *J. Fluid Mech.*, **161**, 313–27.

Barr, D.I.H. (1967). Densimetric exchange flow in rectangular channels. III Large scale experiments. *Houille Blanche*, **6**, 619–31.

Barr, D.I.H. & Hassan, A.M.H. (1963). Density exchange and flow in rectangular channels. II. Some observations of the structure of lock exchange flow. *Houille Blanche*, **7**, 757–66.

Beghin, P., Hopfinger, E.J. & Britter, R.E. (1981). Gravitational convection from instantaneous sources on inclined boundaries. *J. Fluid Mech.*, **107**, 407–22.

Benjamin, T.B. (1968). Gravity currents and related phenomena. *J. Fluid Mech.*, **31**, 209–43.

Britter, R.E. (1979). The spread of a negatively buoyant plume in a calm environment. *Atmos. Environ.*, **13**, 1241–7.

Britter, J.E. & Linden, P.F. (1980). The motion of the front of a gravity current travelling down an incline. *J. Fluid Mech.*, **99**, 531–43.

Chen, J-C, & List, E.J. (1976). Spreading of buoyant discharges. *Proc. 1st CHMIT Seminar on Turbulent Buoyant Convection, Dubrovnik*, pp. 171–182.

Fay, J. (1969). The spread of oil slicks on a calm sea. In *Oil on the Sea*, ed. D.P. Hoult, pp. 43–63. Plenum.

Georgeson, E.M.H. (1942). The free streaming of gases in sloping channels. *Proc. R. Soc., London, Ser. A*, **180**, 484–93.

Gröbelbauer, K.P., Fannelop, T.K. & Britter, R.E. (1993). The propagation of intrusion fronts of high density ratios. *J. Fluid Mech.*, **250**, 669–87.

Henderson, F.M. (1966). *Open Channel Flow*. New York: Macmillan.

Hopfinger, E.J. & Tochon-Danguy, J.C. (1977). A model study of powder-snow avalanches. *Glaciology*, **19**, 343–56.

Hoult, D.P. (1972) Oil spreading on the sea. *Ann. Rev. Fluid Mech.*, **4**, 341–68.

Koh, R.C.Y. (1971). Two-dimensional surface warm jets. *J. Hydraul. Div. ASCE*, **97** (HY6), 819–36.

Linden, P.F. & Simpson, J.E. (1985). Buoyancy driven flow through an open door. *Air Infiltration Rev.*, **6**, 4–5.

Middleton, G.V. (1966). Experiments on density and turbidity currents. I. Motion of the head. *Can. J. Earth Sci.*, **3**, 523–46.

O'Brien, M.P. & Cherno, J. (1934). Model law for motion of salt water through fresh. *Trans. Am. Soc. Civil Eng.*, **99**, 576–94.

Penny, W.G. & Thornhill, C.K. (1952). The dispersion under gravity of a column of fluid supported on a rigid horizontal plane. *Proc. R. Soc. London, Ser. A*, **244**, 283–311.

Rottman, J.W. & Simpson, J.E. (1983). Gravity currents produced by instantaneous releases of a heavy fluid in a rectangular channel. *J. Fluid Mech.*, **135**, 95–110.

Rottman, J.W. & Simpson, J.E. (1984). The initial development of gravity currents from fixed-volume releases of heavy fluids. *IUTAM Symposium, Delft, 1983*, ed. G. Ooms & H. Tennekes. pp. 347–59. Berlin, Heidelberg: Springer-Verlag.

Simpson, J.E. & Britter, R.E. (1979). The dynamics of the head of a gravity current advancing over a horizontal surface. *J. Fluid Mech.*, **94**, 477–495.

Thomas, N.H. & Simpson, J.E. (1983). *The Motion of a Gravity Current Along a Free Surface*. Central Electricity Generating Board report.

Wilkinson, D.L. & Wood, I.R. (1972). A rapidly varying flow phenomenon in a two-layer flow. *J. Fluid Mech.*, **47**, 241–56.

Yih, C-S. (1980). *Stratified Flows*. pp. 204–7. New York: Academic Press.

13   Ambient stratification

Gravity currents flowing through stratified surroundings may generate internal waves and also internal bores. The latter form an important mechanism of mass transfer and have many features in common with gravity currents. Many interesting interactions between currents, bores and waves can occur, depending on the speed of the gravity current and its depth relative to the ambient stratification.

13.1 Two-fluid systems

A two-layer system will be considered first in which one fluid of uniform density lies above another of appreciably greater density, with a sharp interface between them. The behaviour of a gravity current which moves along the lower boundary beneath these fluids depends on the nature of the disturbance that its progress creates at the interface between them.

There are several possibilities, and it will be simpler to consider first the effects of moving a solid obstacle, of similar shape to the front of a gravity current, at different speeds through the stratified system. It will then be possible to see how a gravity current in such a system can sometimes generate an internal bore moving ahead of it.

13.1.1 Internal bore formation

The types of disturbance which may be generated at the interface by a moving obstacle are displayed in figure 13.1. This shows four different flow patterns which can result from a start-up after a state of rest. They are defined in terms of the two non-dimensional quantities $F_0 = U/(g'h_0)^{\frac{1}{2}}$ and $H = d_0/h_0$, representing the speed and height of the obstacle (h_0 and d_0 are the heights of the fluid interface and the obstacle, respectively). The results apply strictly to an upper fluid of infinite depth.

On the left of the inset graph in figure 13.1 are found flows (a) and (d) involving obstacles the depth of which is small compared with the depth of the lower fluid. The effect produced by these smaller obstacles consists only of a simple displacement of the interface either upwards or downwards, according to whether the flow is supercritical or subcritical. That is according to whether its velocity, U, is greater or less than $(g'h_0)^{\frac{1}{2}}$, which represents the maximum speed of small internal waves on the interface. Another way of stating this is 'whether the internal Froude number $U/(g'h_0)^{\frac{1}{2}}$ is greater or less than 1'.

When the height of the obstacle approaches the depth of the layer (b) or exceeds it (c) an upstream influence in the form of a hydraulic jump appears for an increasing range of speeds. Partial blocking may exist, in which the flow above the obstacle is reduced and some of the fluid is forced to move ahead of the obstacle. Blocking of a flow past a stationary obstacle is usually defined as the stagnation of a layer of the fluid leading upstream from the obstacle. So, in the case of a moving obstacle, blocking is taken to mean the movement of a region of fluid upstream at the speed of the obstacle. This is not synonymous with the presence or absence of the upstream influence of the obstacle. In a stratified flow it is possible for the part of the fluid near the bottom to be blocked, while the top part still flows.

In all these cases, the partial or complete blocking results in the generation of a hydraulic jump in the form of an internal bore travelling ahead of the

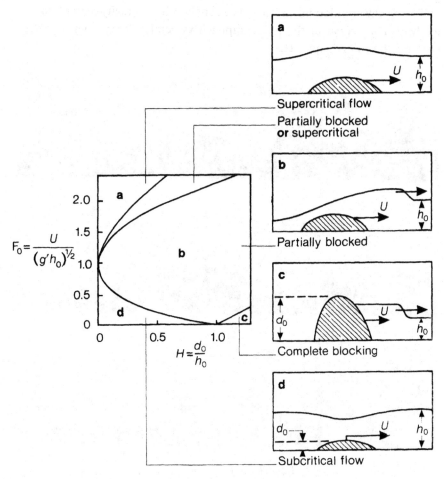

Figure 13.1 Four types of disturbance which can be generated by a moving obstacle at the interface between two fluids.

moving obstacle. Some of the boundaries in figure 13.1 have been verified experimentally (Long, 1970), and experiments on the formation and behaviour of internal bores have been carried out by Wood & Simpson (1984).

The concept of an internal bore has already been introduced in Chapter 1, introducing the important fact of the energy loss which must occur, either by radiation of waves or in turbulence. The character of the bore depends on its strength, the ratio h_1/h_0. The three types of bore produced in the laboratory are shown in figure 13.2. When the strength, h_1/h_0, lies between 1 and 2 (figure 13.2a), the bore has a smooth undular form. The undulations are produced over the obstacle one after another and travel upstream at the bore speed: little or no mixing occurs between the two fluids. When h_1/h_0 lies between 2 and 4 (figure 13.2b), the bore is almost undular, but some mixing occurs on the downstream face of the first undulation, and also at points further downstream, due to shear instability. As h_1/h_0 gets close to 4, this mixing becomes more significant and can be seen on the downstream faces of the first few undulations. When h_1/h_0 is greater than 4 (figure 13.2c), the mixing completely dominates the motion, oblitering any undulations. The bore then appears like a gravity current: this is easily seen by comparing this photograph with several different ones of gravity currents shown in Chapter 11.

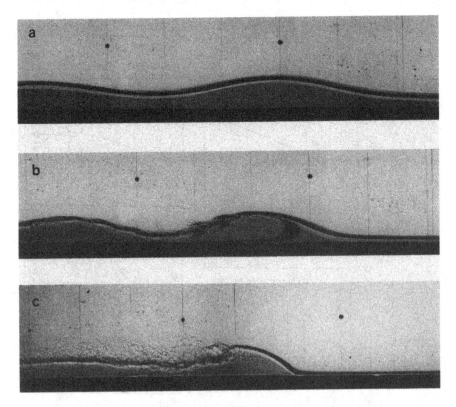

Figure 13.2 Three different forms of internal bore seen at an interface between two fluids. (a) strength (h_1/h_2) between 1 and 2. (b) Strength between 2 and 4. (c) Strength greater than 4.

Theories for bores in two-layer fluids have been developed when there is no mixing between the two fluids (Yih & Guha, 1955; Chu & Baddour, 1977; Wood & Simpson, 1984). The case of greatest interest is that in which the upper layer is very deep compared with the lower layer; the result is then

$$F_0^2 = \frac{1}{2}\frac{h_1}{h_0}\left(1+\frac{h_1}{h_0}\right)$$

This expression is plotted in figure 13.3, together with experimental results (Wood & Simpson, 1984). The agreement is good to about $h_1/h_0=2$, but for larger values of this ratio the mixing between the two fluids reduces the speed below the theoretical value. When h_1/h_0 is greater than 2, the bore speed is quite well predicted by gravity current theory.

Attempts have been made to measure the wavelength in undular bores. Through the range of bore strengths from 1.5 to 2.5 the ratio of the wavelength, λ, to the depth h_1 of the bore is found to be approximately 10.

13.1.2 Generation of bore by gravity current

If a gravity current moves along the ground disturbing a dense layer of fluid, the flow relative to the head can be either supercritical or subcritical. It is more

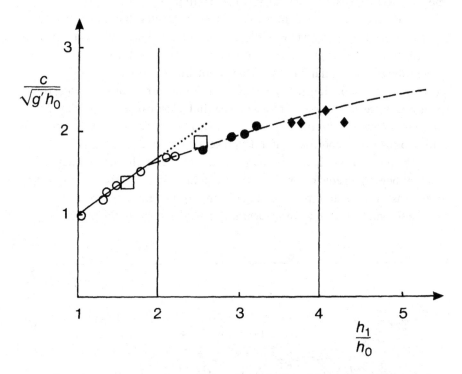

Figure 13.3 Experimental values of the variation of bore speed, c, with strength, h_1/h_0. The speed is non-dimensionalised as the Froude number. $\bigcirc$, $\bullet$ undular bores. $\blacklozenge$ turbulent bores. The straight line to the left represents the undular-bore theory. The curve for $h_1/h_0>2$ is predicted by gravity current theory.

likely, however, that the current will give rise to a hydraulic jump in the form of an internal bore.

The nature of the disturbance produced depends on the relative densities and depths of the current and the layer. The generation of internal bores by gravity currents has been investigated in the laboratory (Rottman & Simpson, 1989), using the arrangement shown in figure 13.4. In this arrangement a gate containing a dense salt solution to a depth D is introduced into a water tank at a given distance from one end of the tank. On the other side of the gate there is a layer of slightly less-dense fluid of depth h_0. Fresh water lies above to a total depth, H, which is greater than D. When the gate is removed the denser fluid begins to move forward as a gravity current along the floor of the tank.

The advancing front of dense fluid acts in a similar way to the solid obstacle already considered, but there are some important differences. If the gravity current is shallower than the undisturbed lower fluid, then it is possible to generate either a supercritical or a subcritical flow, as described in section 13.1.1. In a supercritical flow the gravity current is moving faster than any disturbance can move forward along the top of the dense layer: the interface between the fluids rises smoothly over the head as the gravity current advances. In a subcritical flow the only disturbance is a small depression in the layer which moves along above the advancing gravity current head.

When the depth, d_0, of the gravity current is greater than that of the layer, h_0, the flow of the dense layer is partially blocked and a bore is generated, which may separate from the gravity current and move ahead of it. The first stage in the generation of an undular bore is the formation of a smooth hump which envelops the head of the gravity current. This hump moves forward along the interface between the two layers, leaving behind it some of the denser fluid of the gravity current. The gravity current is disrupted, but a fresh head soon forms, a fresh bulge forms above this, and the process is repeated.

The photographs in figure 13.5 show three stages in the development of an undular bore by an advancing gravity current. In figure 13.5a the bore is starting to grow as a smooth hump enveloping the head of the gravity current. In figure 13.5b a second front and wave are beginning to take shape. In the third view,

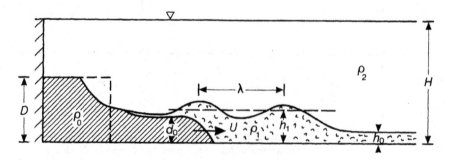

Figure 13.4 Experimental generation of an internal bore by an advancing gravity current travelling to the left.

figure 13.5c, some of the dense fluid beneath the first wave of the bore is starting to be left behind and a third bulge is starting to develop. Soon, all this dense fluid will be left behind and the first wave of the bore will be completely separated from the gravity current. The process will continue as each successive wave separates from the gravity current.

Experimental results in figure 13.6 show the speed of experimental bores generated by gravity currents of various sizes. The sizes of the gravity currents which were forcing the bores are given in terms of the ratio of the depth of the gravity current, d_0 (which is approximately half D) to that of the undisturbed layer, h_0. The speeds, U, are non-dimensionalised as a Froude number, $U/(g'h_0)^{\frac{1}{2}}$, based on g', the reduced gravity, and h_0, the depth of the layer.

The solid curves in the graph mark out the areas of disturbances expected to be formed by the movement of a solid obstacle, as shown in figure 13.1. Additional dashed curves show the expected boundaries of bores of strength 2 and 4. Super- and subcritical flows are plotted as square and circular symbols. As expected, they appear with relatively small gravity currents in regions where the Froude number is greater and less than 1, respectively.

The points with numbers from 1.7 to 3.2 refer to the strengths of bores of undular form, as in figure 13.2a and b. Points in figure 13.6 marked with

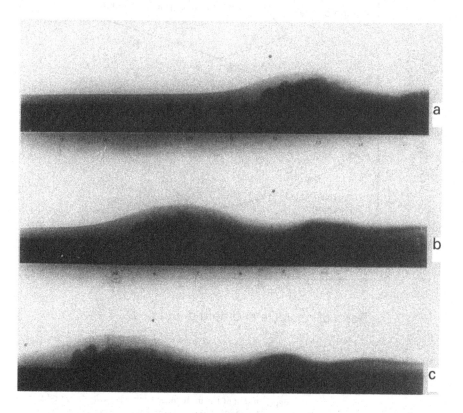

Figure 13.5 Three stages in the generation of an undular bore by an advancing gravity current, travelling to the left.

diamonds refer to turbulent bores with strength greater than 4. If the gravity current has a depth greater than about four times that of the lower layer, this layer is mixed into the gravity current, which then behaves almost as if the lower layer does not exist.

    The area corresponding to 'complete blocking' is blank in figure 13.6 as it does not seem to be possible to achieve this state by a gravity current moving through a layer. The nearest flow situation that can be attained yields bores generated by intrusive gravity currents. These currents do not move along the floor, but travel along the interface between the two layers. Points corresponding to bores generated by intrusive gravity currents lie close to the boundary of the 'complete blocking' zone. Bores formed by intrusive gravity currents will be discussed later in section 13.1.3.2.

13.1.3   Intrusive gravity currents

Both in the ocean and in the atmosphere, many examples of density layering exist which are thought to be caused by intermediate or intrusive gravity

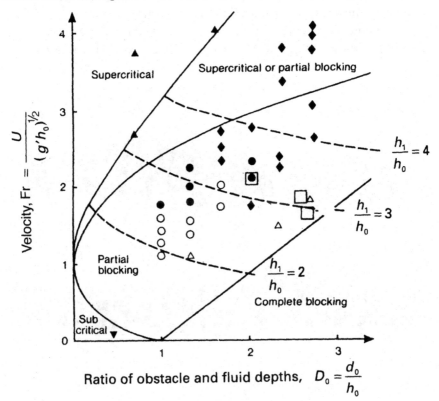

Figure 13.6   Speed of internal bores generated by gravity currents in the laboratory. The speed, U, is non-dimensionalised as the Froude number $U/(g'h_0)^{\frac{1}{2}}$. The size D_0 is the ratio of gravity current depth to that of the undisturbed dense layer, h_0. $\bigcirc$, $\bullet$ undular bores. $\blacklozenge$ turbulent bores. $\square$ atmospheric bores. $\triangle$ bores generated by intrusive gravity currents.

currents. These intrusions may occur after a limited section of two layers of fluid has been thoroughly mixed and released. A gravity current of mixed fluid flows along the interface, and, furthermore, this current can sometimes generate a bore moving ahead of it.

13.1.3.1 *Intrusions along a sharp interface*

A limited class of intrusive gravity currents can be produced very simply in the laboratory. The lower half of a tank is filled with salt solution, and tap water is carefully added to an equal depth above it. A partition is inserted near one end and the two fluids behind it are then stirred until they are completely mixed. When the partition is removed, the mixed fluid collapses and flows as an intrusion along the interface between the two fluids. In a refinement of this apparatus the intrusion is derived from a layer of fluid of intermediate density introduced at an appropriate time during the filling process.

The thickness of the interface between the two fluids, which always increases slowly with time due to molecular diffusion, has significant effects on the nature of the head of the intrusive current. The thickness of the interface between the two fluids in the intrusion shown in figure 13.7 was found to be just less than 20% of the thickness of the gravity current. It has been found experimentally by Britter & Simpson (1981) that this relative thickness can be taken as effectively 'sharp'. The speed of such a symmetrical intrusion varies with fractional depth in much the same way as an inviscid-boundary gravity current.

Intrusive gravity currents which occur in nature may not always have exactly the mean density of the two layers. Asymmetric flows have features of more interest than the symmetrical ones, and these currents are capable of

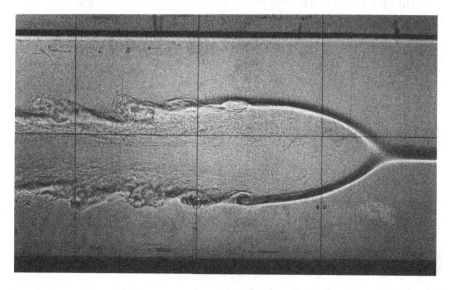

Figure 13.7 Shadowgraph of an intrusive gravity current
running left to right along a sharp interface between two fluids.

generating bores by displacing the level of a sharp interface. A range of internal bores can be produced in the laboratory by replacing the fluid behind the barrier with one of either greater or less density than the mean of the two layers. By altering the relative depth of the layers, a range of internal bores can be produced. Figure 13.8 shows such a bore produced at a thin sharp layer by an intrusive gravity current with density only slightly greater than that of the lower fluid. In this picture the first undulation of the bore has moved about 10 cm ahead of the intrusion, and the second undulation has almost separated from the head of the intrusion.

13.1.3.2 *Intrusion along a thick interface between two fluids*

When the interface between the two fluids is 'thick', the behaviour of an intrusive gravity current is best understood by considering once more the symmetrical case.

In small-scale laboratory experiments it may not be easy to maintain the thickness of the interface at less than 20% of that of the intrusion, as molecular diffusion will continually increase the width of the interface to a thickness which may be significant.

If the interface thickness increases, the speed of advance of a symmetrical intrusion is increased. Figure 13.9 shows some results of experiments to measure the effect of the interface thickness on the speed of a symmetrical intrusion. The current through the thin interface shown in figure 13.7 is represented by the point '□' in the graph. As the relative thickness of the interface increases, the structure of the gravity current alters and when the interface thickness is 70% of the gravity current the head of the current has disappeared and the appearance is that of the reflection of a bore on a surface whose strength is 4 or greater.

The early stages of a bore of this type are shown in the photograph in figure 13.10, in which the bulge forming around the gravity current head almost completely separates the head of the dense fluid from the following flow.

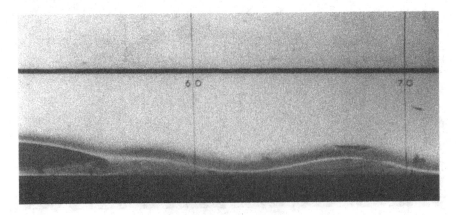

Figure 13.8 Experimental production of a bore by an intrusive gravity current.

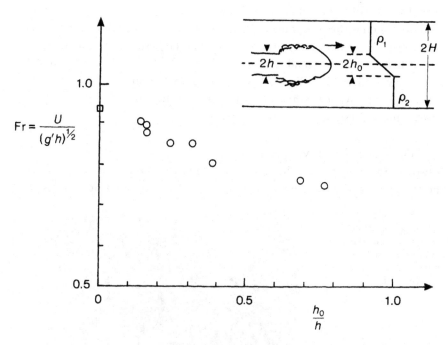

Figure 13.9 Experimental results which show dependence of speed of intrusions on the thickness of the interface between the two fluids.

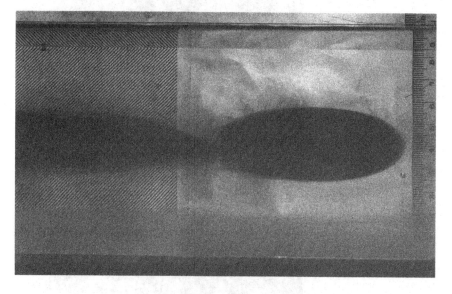

Figure 13.10 Undular intrusive bore observed in a laboratory experiment, in which the head is almost completely separated from the following flow.

Caution is needed in the interpretation of such mixed-layer experiments in which both the lock length and the distance travelled by the current are of limited length. If the lock length is short, then it is possible for the currents above and below, flowing in the opposite direction to the gravity current, to be reflected from the end wall and to catch up the front. If this happens, then the bore may be reduced to merely the appearance of a solitary wave. Also, in a short run of a bore, at the start of its generation, the appearance is that of a simple solitary wave.

Allied experiments made by Maxworthy (1980) in axisymmetric, or radial, flows show similar effects. Figure 13.11 shows two stages in the development of a mixed region advancing through a two-level stratification, in which the development of a series of waves can be seen.

13.1.4 Collision of two gravity currents

If two gravity currents meet they may interact in various ways. In the atmosphere the collision between two mesoscale frontal flows has been observed and it is

Figure 13.11 Sequence of waves generated by an axisymmetric mixed region advancing through a two-layer stratification. (Courtesy of T. Maxworthy.)

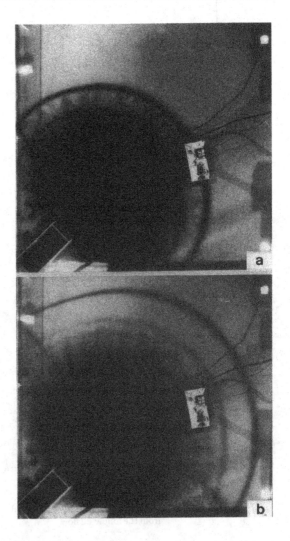

found that the combining of the rising air at the two fronts produces an especially strong zone of convection (Findlater, 1964). There is also evidence, from sedimentary deposits, of the collision of turbidity currents on the ocean bed.

The collision of a gravity current with a vertical wall was described in Chapter 11 where figure 11.16 showed the reflection to be in the form of an undular bore. Only a few studies of the collision between two gravity currents have been published, but both experimental results (Kot & Simpson, 1987) and a numerical study by Clarke (1984) show the main effect to be the emergence of two undular bores in opposite directions.

The results of a collision between two gravity currents of different size and density are shown in the four photographs of figure 13.12. The form of each front soon after the collision, shown in figure 13.12b, much resembles the shape of a gravity current being reflected from a vertical wall. Developing bores can be seen in figure 13.12c and d, moving away from each other. Since the two gravity currents were not equal in height and density the current from the left is reflected as a deep turbulent bore, but that from the right reappears as a weak one, of undular form, possibly only consisting of a solitary wave.

Figure 13.13 gives the results of such an experiment, showing the development of the bores (Figure 13.13a) and the velocities of the approaching fronts with the resultant bores (figure 13.13b). Nearly all the energy is transferred to the two bores, and only small gravity currents continue to travel forwards. If the fluid in each current in the experiments were not distinguished from the other by the use of a dye, it would be easy to make the mistake that the two gravity currents had met and 'crossed', and that each one continued to travel in the same direction as before.

'Crossing fronts' in the atmosphere have been described, in which clear disturbances emerged from a collision and moved at about the same speed and direction as before the meeting (Rider & Simpson, 1968). It is likely, however, that only a small amount of the fluid actually crosses and that these events involve the formation of undular bores. In these atmospheric encounters it is often difficult to distinguish between a bore and a gravity current, as a wide spectrum of phenomena exists between a pure bore and the gravity current which initiated it.

The formation of an atmospheric bore from a collision has been confirmed by Wakimoto & Kingsmill (1995), who examined the meeting of a sea-breeze front and a gust front over central Florida in August 1991. Using doppler radar, satellite imagery, rawinsondes, cloud photography and surface meso-net data they established the three-dimensional structure of an undular bore which resulted from the collision.

13.2 Continuous stratification

A stratification with a constant density gradient can easily be set up in a laboratory using the 'two-tank' method. Dense fluid is pumped slowly into the experimental channel from the first storage-tank, which originally contains fluid of the maximum density required. This tank is continuously replenished by fresh water stirred in from the second tank until the first contains only fresh water.

A fundamental quantity related to the stratification is the Brunt–Väisälä frequency, sometimes simply called the buoyancy frequency. This is the frequency of oscillation of a small particle of the fluid if disturbed a small distance up or downwards. It is usually denoted by N (per second), and is given by $(-g/\rho \cdot \mathrm{d}\rho/\mathrm{d}z)^{\frac{1}{2}}$.

In fluid with a constant density gradient ($\mathrm{d}/\mathrm{d}z$ is constant) the value of the buoyancy frequency is constant with depth throughout the fluid. In this case an

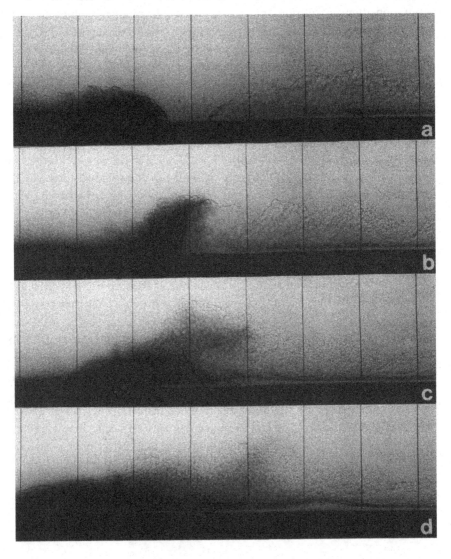

Figure 13.12 Four stages in the collision between two unequal gravity currents.

important relationship exists with the Froude number. Using the notation in figure 13.14, the density gradient $= -d\rho/dz$, so $\Delta\rho/H = -d\rho/dz$.

So $Fr = \dfrac{U}{\left(\dfrac{\Delta\rho}{\rho_0} \cdot gH\right)^{\frac{1}{2}}}$

And since $N = \left[-g/\rho \dfrac{d\rho}{dz} \right]^{\frac{1}{2}}$

it follows that $Fr = U/NH$

a

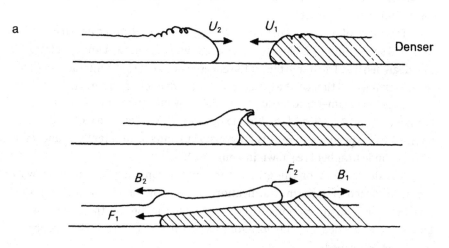

b

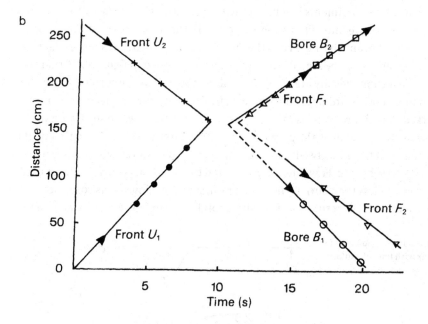

Figure 13.13 Experimental collision of two gravity currents.
(a) Formation of two reflected bores. (b) Velocities of fronts and bores.

Theory and experiments (Long, 1955) have shown that the movement of an obstacle through a uniform stratification can generate several different modes of internal waves. The maximum velocity of disturbances which can move upstream of an obstacle of height H is $c = NH/\pi$, or in other words, for disturbances to move upstream the Froude number, U/NH, must be less than $1/\pi$.

13.2.1 Gravity currents in uniform stratification

This section will consider the passage of a gravity current along a boundary through a medium which is uniformly stratified, or in other words has a constant density gradient.

The speed of a gravity current travelling through a limited depth of stratified surroundings is close to that of a gravity current in a uniform fluid of the mean density. The important differences are related to the interaction with the internal waves that can be generated by the advance of the gravity current.

In the experiments to be described the buoyancy frequency, N, was varied between 1.3 and 2.0 s^{-1}, and the Froude number, U/NH, between 0.05 and 0.4. Some of the experimental results for gravity currents of different size and values of the Froude number are shown in figure 13.15.

The shaded areas on this graph represent the zones in which internal waves can be produced in uniform stratification by a moving solid obstacle (Long, 1955). They show that no interaction is to be expected to the right of the dashed line, and the three shaded areas enclose zones where first, second and third order waves can be formed.

In these experiments in which gravity currents took the place of the solid obstacles, no wave interactions were observed in the results marked with open circles in the graph. In these runs the Froude number, U/NH, was just greater than $1/\pi$, and it can be seen from the example illustrated in figure 13.16 that the lines of constant density are merely displaced upwards as the current arrives and then move downwards after the passage of the gravity current head. This run, marked 'A' in figure 13.15, is in a zone where no disturbance can separate and move away upstream of the gravity current and the effect is similar to that of the supercritical flows in a two-layer medium, described in section 13.1.2.

However, when Fr was less than $1/\pi$ and the fractional depth of the gravity current was less than 20% of the total depth, some striking effects could be observed. In all the results marked with a solid circle in figure 13.15, waves were

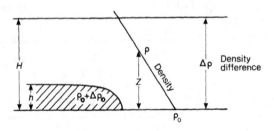

Figure 13.14 Notation used for a tank with fluid of uniform density gradient.

set up by the head of the advancing gravity current. These waves affected the form of the current behind the head in a rhythmical manner. The dense fluid in the original head was cut off from the flow, which formed a second head. The process was repeated and later a third new front appeared. The dissolution of each head was followed by the development of a new one to follow it. A photograph of one such experiment in progress is shown in figure 13.17, in which the rapidly forming second front is already well formed behind the original front, which has been cut off and is dissolving away.

The experiment in this photograph corresponds to the point marked 'B' in

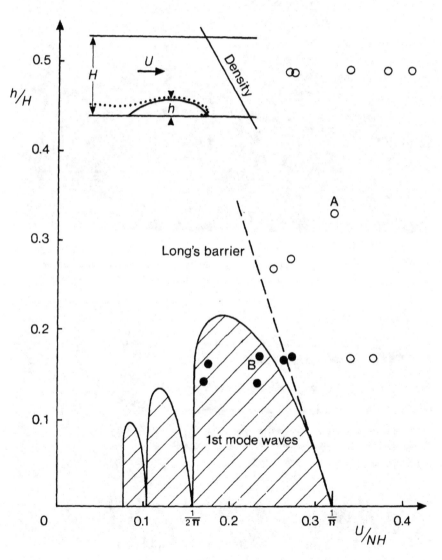

Figure 13.15 Experimental results from gravity currents of different size and Froude number, advancing through fluid of uniform density gradient. Closed circles show values where interaction was seen between the gravity current and the waves it produced.

figure 13.15. The point is one of a set of six results, plotted as solid circles in figure 13.15, which are all close to the curve enclosing values in which first-mode waves are forecast to be formed by the passage of a corresponding solid obstacle.

Similar effects have been seen at a sufficiently low Froude number in experiments in which a gravity current flows along the surface above a uniformly stratified fluid. In the ocean such conditions are likely to occur near the surface in brackish water in the entrance to a fjord, and we discuss some observations in Chapter 15.

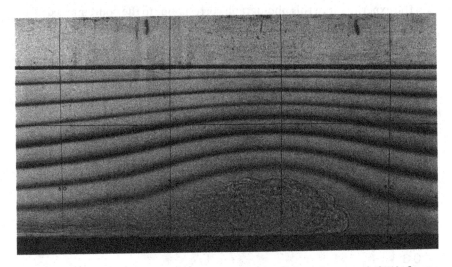

Figure 13.16 The flow seen in the experiment marked 'A' in figure 13.15. The Froude number is greater than $1/\pi$, and the flow is supercritical. The dark lines are lines of constant density.

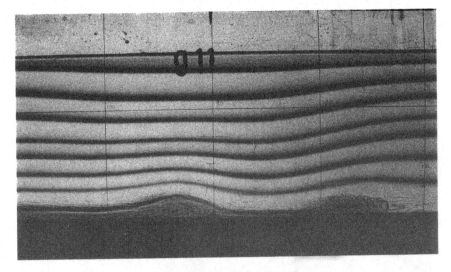

Figure 13.17 One stage in the experiment marked 'B' in figure 13.15. The front generated first-mode waves in the stratified fluid, which interacted with the gravity current flow. The original front is dissolving away and a second front has started to form.

13.2.2 Intrusions into constant stratification

Intrusive gravity currents are produced in the environment in surroundings of constant density gradient by the collapse of a region which has become mixed by turbulence; this mixed fluid then moves forward as a gravity current, flowing at a height appropriate to its density. These intrusions may be responsible for the formation of internal waves whose properties have been examined in laboratory experiments.

The complete process of collapse has been divided into three stages. After the initial stage of rapid collapse, the principal stage follows, which is also controlled by gravitational forces. In the final stage the flow is controlled by viscous forces and is in the form of a long thin wedge.

Internal waves similar to those formed during the early stage of the collapse can conveniently be generated by an oscillating plunger (Wu, 1969). The pattern of these internal waves can be represented by moving rays connecting either wave crests or troughs. At first these rays decrease their slopes, but a simple steady-state pattern is formed in the later stages. Figure 13.18 represents the first two stages and the wave pattern. It has been shown that the angle θ of the rays is given by

$$\theta = +\sin^{-1}(f/N)$$

where f is the frequency of the plunger oscillations and N is the buoyancy frequency.

Examination of the progress of gravity currents moving through uniform stratification, both as two-dimensional and as three-dimensional radial flows, shows the pulsing flow already described for boundary gravity currents by Amen & Maxworthy (1980). As noted above, this behaviour is a result of strong interaction between the wave system and the gravity current in which the current is left behind as it is passed by successive waves.

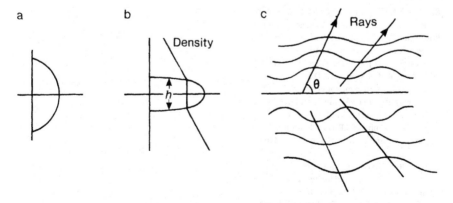

Figure 13.18 Stages in the generation of waves by an intrusion into uniformly stratified fluid.

The conditions in these experiments closely resemble those found both in the atmosphere and in the ocean, which we have already examined in Chapters 5 and 7.

Bibliography

Amen, R. & Maxworthy, T. (1980). The gravitational collapse of a mixed region into a linearly stratified fluid. *J. Fluid Mech.*, **96**, 65–80.

Britter, R.E. & Simpson, J.E. (1981). A note on the structure of an intrusive gravity current. *J. Fluid Mech.*, **112**, 459–66.

Chu, V.H. & Baddour, R.E. (1977). Surges, waves and mixing in two-layer density stratified flow. *17th Congress, International Association of Hydraulics Research, Baden-Baden*, **1**, 303–10.

Clarke, R.H. (1984). Colliding sea breezes and the creation of internal atmospheric waves: a numerical model. *Aust. Meteorol. Mag.*, **32**, 207–26.

Findlater, J. (1964). The sea-breeze and inland convection: an example of their interrelation. *Meteorol. Mag.*, **93**, 82–9.

Kot. S.C. & Simpson, J.E. (1987). Laboratory experiments on two crossing flows. In *Proc. 1st Conf. Fluid. Mech. Beijing, 1987*, pp. 731–6. Beijing University Press.

Long, R.R. (1955). Some aspects of the flow of stratified fluids. III. Continuous density gradients. *Tellus*, **3**, 341–57.

Long, R.R. (1970). Blocking effects in flow over obstacles. *Tellus*, **22**, 471–80.

Maxworthy, T. (1980). On the formation of nonlinear internal waves from the gravitational collapse of mixed regions in two and three dimensions. *J. Fluid Mech.*, **96**, 47–64.

Rider, G.C. & Simpson, J.E. (1968). Two crossing fronts on radar. *Meteorol. Mag.*, **97**, 24–30.

Rottman, J.W. & Simpson, J.E. (1989). Internal bores in the atmosphere and in the laboratory. *Q. J. R. Meteorol. Soc.*, **115**, 941–63.

Wakimoto, R.M. & Kingsmill, D.E. (1995). Structure of an atmospheric undular bore generated from colliding boundaries during CAPE. *Mon. Wea. Rev.*, **123**, 1374–93.

Wood, I.R. & Simpson, J.E. (1984). Jumps in layered miscible fluids. *J. Fluid Mech.*, **140**, 329–42.

Wu, J. (1969). Mixed region collapse with internal wave generation in a density-stratified medium. *J. Fluid Mech.*, **35**, 531–44.

Yih, C.S. & Guha, C.R. (1955). Hydraulic jump in a fluid system of two layers. *Tellus*, **7**, 358–66.

14   Ambient turbulence

An example of a turbulent flow can be experienced by standing out of doors in a wind. Although the wind strength may be indicated for example as 10 m s^{-1}, rapid variations of several metres per second can be felt. These arrive as gusts then die away and the process is continually repeated. The flow can be pictured as a mixture or a collection of eddies, all of different sizes and strengths, and the speed of 10 m s^{-1} can only represent some kind of averaged value.

When the front of a gravity current moves in calm surroundings, most of the mixing occurs in the region of the head and this results in the laying down of a stable layer above the following current. In cases when the level of ambient turbulence is small it usually has only a small effect on this mixing process, but if the continued spread of a finite amount of dense fluid into a turbulent environment is considered this may no longer be true. Although at the start the normal processes dominate, eventually as the density becomes smaller and the velocity of the front is sufficiently reduced, the effect of external turbulence may become appreciable. When the current has reached this stage the form of the leading edge begins to change and the depth of the current, which had been decreasing, may begin to increase.

14.1 Turbulence experiments

Many investigations have used the theoretical concept of turbulence in which the statistical properties do not vary with position and have no preferred direction. This is known as 'homogeneous isotropic turbulence'.

A typical grid used for generating turbulence in a wind or water tunnel is shown in figure 14.1. Turbulent energy is produced in the flow past the grid; this becomes almost isotropic and homogeneous with distance downstream.

Another way of producing turbulence is by regular oscillations of this kind of grid. This method has been used to generate turbulence in water tanks, especially to examine the effects of turbulence on stratified fluids. A measure of the turbulent intensity produced by the oscillation of such a grid at a specified distance can be given in terms of the values of representative velocity fluctuations U' and length fluctuations L'. Extensive experimental measurements have been made of such turbulent properties produced by grids of different sizes and proportions, oscillating at various frequencies (Turner, 1968; Hopfinger & Toly, 1976).

14.1.1 Experiments using oscillating grids

An oscillating grid was used in a series of experiments by Thomas & Simpson (1985) to maintain and enhance the shear turbulence in a streaming flow over a stationary saline gravity current. As shown in figure 14.2, the moving floor passed beneath a fixed-plate working section. The belt travelled at the mean-depth speed of the flow, U, in order to minimise mean velocity shear in the stream. The turbulence generated by an oscillating grid overhead propagated downwards, whilst being advected by the stream. Unrepresentative mixing of the entering current was suppressed by the shield plate.

Salt solution was continuously supplied at the downstream end of the gravity current in order to compensate for fluid transported into the stream. The supply was adjusted until the front of the gravity current stabilised at the

Figure 14.1 Typical turbulence-generating grid.

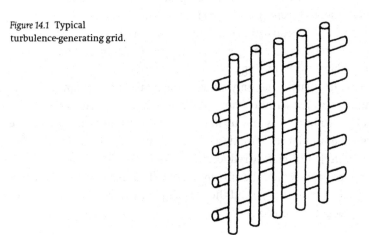

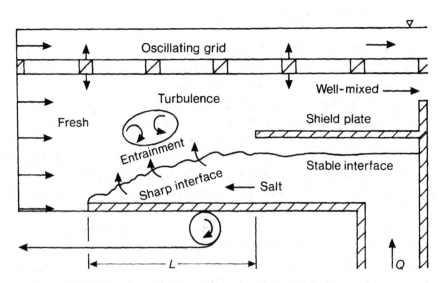

Figure 14.2 Apparatus for measuring the effect of turbulence on a gravity current in a steady state.

leading edge of the fixed floor. Using this apparatus it was possible to maintain turbulent intensities U'/U at any level from zero (the normal gravity current) to about ten or more, the kind of value found in tidal flows. At these large intensities the frontal slope appeared as a wedge, with a fluctuating well-defined interface separating the undiluted current flow and the turbulent surroundings. The current, maintained by a continuous flow from behind, and protected from turbulence, was eroded, not diffused. Figure 14.3 shows the non-dimensional mixing rate appropriate for gravity currents in non-turbulent streams plotted against turbulent intensity U'/U. The measurements approach normal gravity current results (Britter & Simpson, 1978) for $U'/U < 0.5\%$, to the left of line A in the diagram. The mixing flux becomes independent of stream speed for $U'/U > 50\%$, to the right of line B, as shown by the approach to a gradient of three in the log–log plot. Here the mixing rate is two to four times larger than the non-turbulent values.

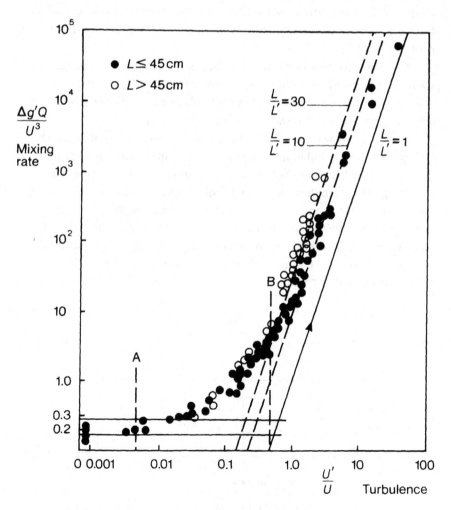

Figure 14.3 The mixing rate in a gravity current subjected to different intensities of turbulence.

In these experiments the process seems more akin to erosion rather than turbulent diffusion, and the results were found to be consistent with those obtained by Turner (1968) in oscillating-grid experiments in a static tank.

14.1.2 Dissipation of fronts

Laboratory experiments on the effects of turbulence have been related to the spreading of salinity in estuaries (Harleman & Ippen, 1960). A long rectangular channel was used with inflow of fresh water at one end and a portion at the opposite end in which a constant salinity was maintained. The mixing effect of the tide was simulated by turbulence generated by oscillating screens at various amplitudes and frequencies.

A turbulent diffusion coefficient K was determined with zero density difference, dye being used as tracer, using a one-dimensional convective diffusion equation. A larger, gross, diffusion coefficient K' is measured in the presence of a density difference induced gravitational convective circulation. The ratio K'/K was taken as a measure of the degree of mixing or stratification in an estuary.

Rising bubbles have been used as an alternative to grids or screens to generate turbulence in laboratory experiments. In particular, the effect on front formation and dissipation in buoyancy-driven flows has been examined in turbulence produced in this way by Linden & Simpson (1986).

Lock exchange was used to establish a saline gravity current flow, which was kept turbulent by bubbling air from a manifold in the base of the tank, as shown in figure 14.4. When the gate is opened a gravity current begins to flow, but mixing by the turbulence starts to transfer momentum between the two counter-flowing layers and reduce the vertical stratification. Eventually, a vertically mixed state is established in which there is an expanding region of horizontal density gradient, centred on the original position of the partition. Figure 14.5 illustrates the stages in the change from a gravity current front to such a vertically mixed region.

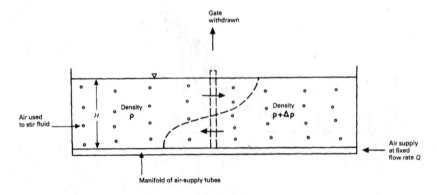

Figure 14.4 Apparatus for releasing a gravity current into liquid made turbulent by ascending air bubbles.

The advance of the gravity current front is plotted against the elapsed time in non-dimensional form in figure 14.6 for a number of different values of g', the original reduced gravity, and H, the depth of the tank. In an undisturbed environment the current flows at the constant speed shown by the broken line. In the initial stages the data follow this line but the turbulence eventually retards the current. The departure from the line at long

Figure 14.5 Three stages in which a gravity current changes into a vertically mixed area in a turbulent water tank.

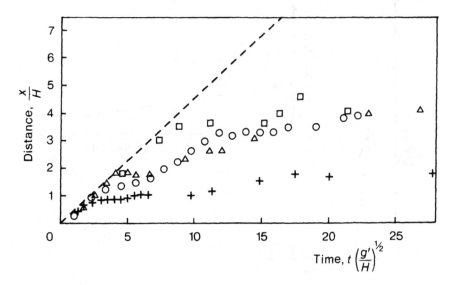

Figure 14.6 The advance of the front of a gravity current with time in turbulent surroundings. In an undisturbed environment a gravity current travels at the rate shown by the broken line. Values of g': $+$, 3.1; $\triangle$, 6.7; $\bigcirc$, 12.6; $\square$, 15.6 cm s^{-2}.

times increases with increasing g' and H, and it is possible to determine an eddy 'diffusivity' K for the longitudinal transport.

It should be emphasised that in the absence of turbulence the longitudinal density flux is at its maximum and the presence of the turbulence decreases the flux.

14.1.3 Formation of fronts

A sharp front can be formed from a horizontal density gradient when the turbulence is reduced. This situation occurs in the atmosphere, where the

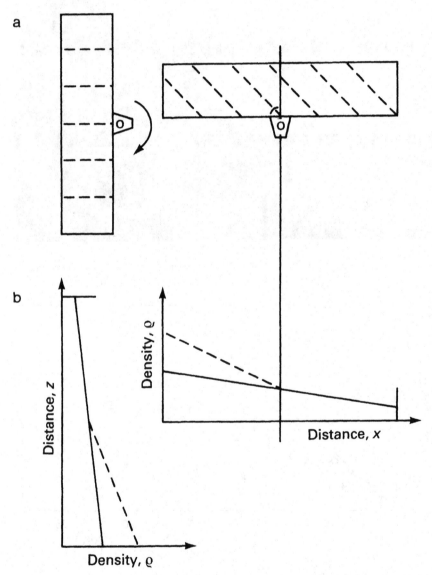

Figure 14.7 Tilting tank apparatus used to examine the formation of a gravity current front. (a) Experimental arrangement. (b) Density gradient: ——, uniform; – – – –, gradient change at centre of tank.

formation of a sea-breeze front may be inhibited until the latter part of the day. As the level of convective turbulence dies down a sharp sea-breeze front may develop.

The formation of a sharp front from a horizontal density gradient was investigated by Linden & Simpson (1989) using the apparatus shown schematically in figure 14.7a. In this 'tilting-tank' experiment a vertical density gradient with horizontal density contours marked by dye was set up, using the two-tank method, in a vertical tank. The tank is then rapidly rotated to a horizontal position and the resulting flows observed as the density markers return to the horizontal position.

When the density gradient is uniform, as marked by the solid line in figure 14.7, the lines show a steady return to the horizontal, except for some disturbances near the ends of the tank. However, where the density gradient is steepened in one half of the tank (but with no density step), as shown by the dotted line, a steadily sharpening front develops.

Figure 14.8 shows two stages in the development of one of these experimental fronts from a non-uniform horizontal density gradient. The converging density lines can be seen in figure 14.8a and a sharp shadowgraph line has appeared at the developing front. In figure 14.8b, as many as four 'density lines' have converged at the front, which is now well developed with turbulent mixing at the leading edge.

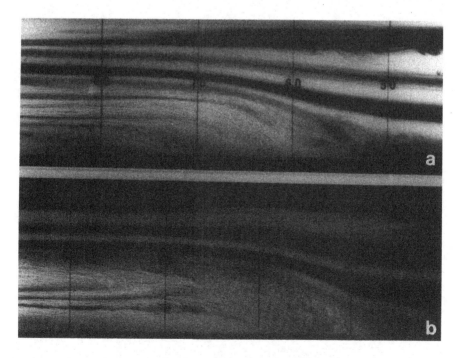

Figure 14.8 Two stages in the development of a gravity current front in a non-linear density gradient. In (a) a sharp line appears at the front; in (b) a fully developed turbulent front has formed.

Bibliography

Britter, R.E. & Simpson, J.E. (1978). Experiments on the dynamics of a gravity current head. *J. Fluid Mech.*, **88**, 223–40.

Harleman, D.R.F. & Ippen, A.T. (1960). The turbulent diffusion and convection of saline water in an idealised estuary. *Int. Assoc. Sci. Hydrol. Commission of Surface Waters*, publ. no. 51, pp. 362–78.

Hopfinger, E.J. & Toly, J.A. (1976). Spatially decaying turbulence and its relation to mixing across density interfaces. *J. Fluid Mech.*, **78**, 155–75.

Linden, P.F. & Simpson, J.E. (1986). Gravity-driven flows in a turbulent fluid. *J. Fluid Mech.*, **172**, 481–97.

Linden, P.F. & Simpson, J.E. (1989). Frontogenesis in a fluid with horizontal density gradients. *J. Fluid Mech.*, **202**, 1–16.

Thomas, N.H. & Simpson, J.E. (1985). Mixing of gravity currents in turbulent surroundings: laboratory studies and modelling implications. In *Turbulence and Diffusion in Stable Environments*, ed. J.C.R. Hunt, pp. 61–95. Oxford: Clarendon Press.

Turner, J.S. (1968). The influence of molecular diffusivity on turbulent entrainment across a density interface. *J. Fluid Mech.*, **33**, 39–56.

15   Viscous gravity currents

Chapter 11 illustrated the change in form of the head of a gravity current
for different values of the Reynolds number. The behaviour of many gravity
currents in the environment appears to be independent of viscous effects, since
no variation in the properties of the flow is apparent with changes in Reynolds
number. For example, in the case of thunderstorm outflows in the atmosphere,
although there must be small-scale viscous effects near the ground, it is known
that their effect on the general dynamics of the flow is negligible provided the
Reynolds number, UL/ν, is greater than a certain value. The magnitude of this
critical value has been shown in laboratory experiments on gravity currents by
Simpson & Britter (1979) to be between 500 and 1000, so thunderstorm outflows,
with a Reynolds number of over 10^6, are well beyond this limit.

However, there are many large-scale examples, such as lava flows and
currents of debris or mud, in which viscous forces play a large part in the
dynamics of the flow. Useful insight into the behaviour of this type of viscous
flow can also be obtained from laboratory experiments.

Low-Reynolds-number experiments can be carried out with flows of
salt solution in water, provided that both the size and the density differences
are kept very small. One of these low, slow currents is pictured in figure 15.1;
the speed of this flow is about half a centimetre per second and the depth is
half a centimetre, so the Reynolds number Uh/ν of the flow is 0.5×0.5/0.01, or
25. This flow is very smooth and the elevated head of the current is small; for

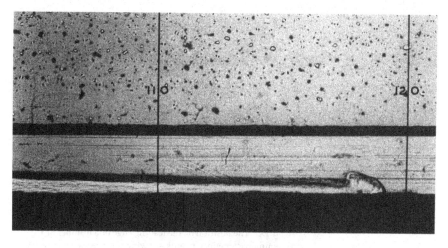

Figure 15.1 A saline gravity current with Reynolds number of 25.

values of Re less than 10 the head no longer exists and the flow is a smooth wedge.

For experiments with much lower Reynolds number a viscous material like treacle has been used, and there are a number of silicon oils available to cover the range of viscosities required.

15.1 Inertial and viscous regimes

In some environmental flows, for example in the spread of oil on the surface of the sea, although viscous effects may not be significant at first, they become more important in the later stages when the layer of oil becomes thin and moves more slowly. A stage is finally reached when the flow is controlled mainly by viscous forces. (In the case of spreading oil, surface tension forces may also become important in the later stages.)

From dimensional analysis it can be shown that

$$\text{inertial force/viscous force} = \frac{\rho D^2 h/t^2}{\rho \nu D^2/th} = \frac{(h/t)h}{\nu}$$

which can be written Uh/ν, the form we have already described for the Reynolds number of the flow. So we see that the Reynolds number, Uh/ν, can be expressed as the ratio of the inertial to the viscous forces determining the flow.

Figure 15.2 shows results of a series of experiments in which a fixed flux of salt water was released into an axisymmetric tank containing much deeper fresh water (Britter, 1979). The expected temporal variation of the position of the leading edge of the flow, normalised by

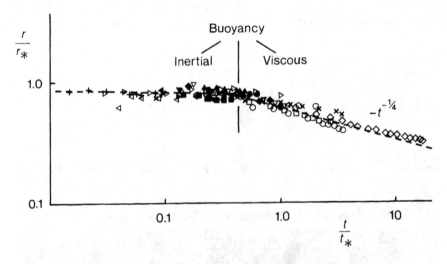

Figure 15.2 The position of the leading edge of a number of axisymmetric gravity currents, showing the passage from the inertial to the viscous regimes. (Courtesy of R.E. Britter.)

$$r_* = (Q_1 g')^{\frac{1}{4}} t^{\frac{3}{4}}$$

is plotted against t/t_*, where the time is normalised with the viscous time-scale (Chen & List, 1976).

$$t_* = Q^{\frac{1}{4}} / (\nu g')^{\frac{1}{2}}$$

The experimental results lie on a single curve, and demonstrate clearly the shift from the buoyancy–inertial regime to the buoyancy–viscous regime as each flow proceeds.

15.2 The spreading of viscous gravity currents

The spread of both two-dimensional and axisymmetric currents over a rigid horizontal surface has been studied theoretically and the results compared with laboratory experiments (Huppert, 1982a; Didden & Maxworthy, 1982). Except possibly in the early collapse stages, the length of the current will greatly exceed its thickness, which allows for the boundary-layer approximation common to lubrication theory to be used in the analysis. The effect on the gravity current of the motion of the upper fluid can be entirely taken into account by applying the boundary condition of zero stress at the upper surface of the current. The current is thus identical with one propagating with a free surface beneath a fluid of negligible inertia.

The spread of viscous gravity currents is found to be as follows:

Radial: $\qquad rN = 0.623 (g' Q^3 / \nu)^{\frac{1}{8}} t^{\frac{1}{8}}$

Two-dimensional: $\qquad xN = 0.804 (g' Q^3 / \nu)^{\frac{1}{5}} t^{\frac{4}{5}}$

One important result from this theoretical treatment is that conditions at the front of a viscous current play no part in determining its motion or shape. This lack of influence of the front will only be true if the Reynolds number is low enough and if the effects of surface tension are small.

This is different from high-Reynolds-number gravity currents, which are totally controlled by conditions at the front. As has been described in this book, many theoretical and experimental studies have been devoted to determining the controlling condition, or Froude number, at the front when the Reynolds number is very great.

15.3 Viscous flows on slopes

If viscous fluid is released on a horizontal surface it takes up a circular plan form which is stable to small disturbances. However, if fluid is released on a sloping surface – for example, some liquid detergent on a sloping plate – a very different plan form occurs. If a broad band of viscous fluid, uniform in depth across the slope, is released it first moves steadily downstream and then breaks up into a series of waves of increasing amplitude. An example is shown in figure 15.3. It has been shown that the instability cannot be explained by viscous effects alone and that surface tension must be taken into account (Huppert, 1982b).

Experiments using silicone oils indicate that the wavelength of the instability is independent of the coefficient of viscosity. Other experiments, with two fluids of comparable viscosity but different surface tension, show that the wavelength is a function of the surface tension.

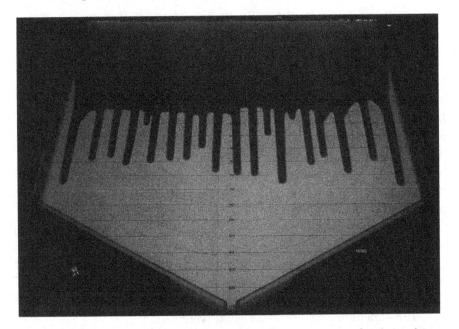

Figure 15.3 Instability at the front of a viscous flow down a slope. (Courtesy of H.E. Huppert.)

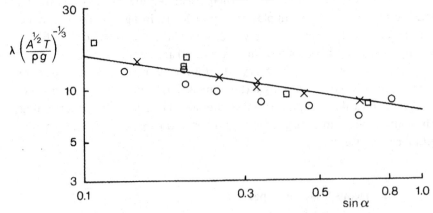

Figure 15.4 Wavelength (λ) of the instability at the front of a viscous flow down a slope of angle α, from three experiments.

The graph in figure 15.4 shows a plot of experimental results (Huppert, 1982a) of the wavelength, λ, normalised with respect to $(A^{\frac{1}{2}}T/\rho g)^{\frac{1}{3}}$ as a function of the slope angle, α. The data collapse onto a single curve and are well represented by

$$\lambda = 7.5(A^{\frac{1}{2}}T/\rho g \sin\alpha)^{\frac{1}{3}}.$$

Bibliography

Britter, R.E. (1979). The spread of a negatively buoyant plume in a calm environment. *Atmos. Environ.*, **13**, 1241–7.

Chen, J-C. & List, E.J. (1976). Spreading of buoyant discharges. *Proc. 1st CMIT Seminar on Turbulent Buoyant Convection, Dubrovnik*, pp. 171–81.

Didden, N. & Maxworthy, T. (1982). The viscous spreading of plane and axisymmetric gravity currents. *J. Fluid Mech.*, **121**, 27–42.

Huppert, H.E. (1982a). The propagation of two-dimensional and axisymmetric viscous gravity currents over a rigid horizontal surface. *J. Fluid Mech.*, **121**, 43–58.

Huppert, H.E. (1982b). Flow and instability of a viscous current down a slope. *Nature*, **300**, 427–9.

Simpson, J.E. & Britter, R.E. (1979). The dynamics of the head of a gravity current advancing over a horizontal surface. *J. Fluid Mech.*, **94**, 477–495.

16 Suspension flows

The density difference in gravity currents can be caused by thermal effects or as the result of dissolved material. Another very important type of gravity current is that in which the density difference is caused by *suspended* material. These suspension gravity currents are of especial interest to the geologist, the oceanographer and the civil engineer. There is a wide range of such flows, which include streams of muddy or turbid water, powder-snow avalanches, and flowing material from some kinds of volcanic eruptions.

If the suspended material which causes the increased density of the fluid consists of very fine particles, then the rate of fall-out may be very small. For a considerable time the behaviour of this suspension current will be similar to that of a gravity current caused by dissolved material.

An example of such a suspension current can often be seen in the atmosphere when a large building is being demolished. As the building collapses, air is displaced, together with large amounts of very fine suspended dust particles; this results in a suspension gravity current moving away from the building . A good example is shown in figure 16.1, which is a view of one of these

Figure 16.1 A gravity current formed by suspended dust from a partly demolished building. (Photo by Associated Press.)

gravity currents moving towards the observer. (Part of the building can still be seen standing behind the current, since in this photograph it has only collapsed to half its original height.)

16.1 Turbidity currents

Suspension currents are also common in water. They can easily be demonstrated in clear, calm pond water by using a stick to stir the mud at the bottom of a pond. A turbidity current of muddy water will spread along the bottom of the pond away from the disturbance. Similar muddy currents can be seen when a clear stream is joined by a muddy river.

The mechanism of flows of muddy, or turbid, water was first recognised in the study of sediment-laden streams continuing beneath the surface of larger bodies of water. Figure 16.2 shows a diagram of such a flow of dense turbid water entering a lake or reservoir and starting to flow as a turbidity current beneath the less-dense fresh water. Where a turbid river enters a lake a sharp boundary between clear and muddy water is visible at the surface.

It is known that a turbidity current on the sea bed can erode a deep channel, rather as a flowing river forms a channel. The sediment transported and deposited by turbidity underflows in this way is of great significance in the development of the sea bed and in the formation of sedimentary rocks.

16.1.1 Experiments on turbidity currents

The creation of a controlled laboratory turbidity current involves considerable practical difficulties, in contrast to the relative ease with which saline currents can be created and their density measured.

The first experiments on turbidity currents were described by Kuenen

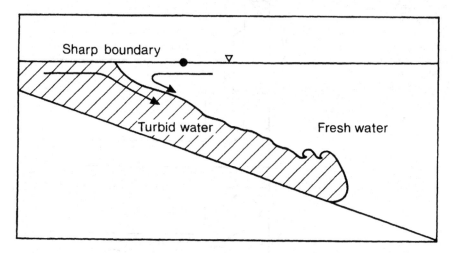

Figure 16.2 A turbidity current flowing into a large volume of fresh water.

(1951), and showed that high-density turbidity currents could be produced in the laboratory. A dense mixture of sand, silt and water was released down a short slope under clear, still water. Although the vertical distance of fall was only about 50 cm, this was sufficient to give an underflow moving at more than 5 cm s^{-1}, which then continued along the full length of about 30 cm of a horizontal channel. A front view of such a laboratory turbidity current has already been shown in the first chapter of the book, in figure 1.3.

From such experiments it was concluded that oceanic turbidity currents falling through a distance of several kilometres could reach the high velocities deduced from deep-sea evidence.

Except in cases where the deposition of the suspended solid material is negligible during the time of the experiment, the simple lock-exchange technique used for saline gravity currents is inadequate. In later experiments mechanisms were devised for introducing the material to be suspended to start a turbidity current in 'running order', (Middleton, 1966; Riddell, 1969). Figure 16.3 shows one such arrangement. The material to be suspended is submerged in the introduction box, and when the gate is removed the material drops and forms a suspension current. This current forms an underflow from which material settles out on the bed of the tank.

Such turbidity currents show most of the features already observed in saline gravity currents (Allen, 1971), for example in the nature of mixing at the head and the relationships between velocity, depth and density difference, both along level floors and down slopes. One notable difference, however, was the threshold for Re-independent flows, beyond which the viscous regime was entered. It was found that this regime was first attained in these flows at appreciably higher Reynolds numbers (Riddell, 1969).

Figure 16.3 Experimental arrangement for production of turbidity currents. (a) Material submerged in introduction tank. (b) Gate removed. (c) Underflowing turbidity current. (After Riddell, 1969.)

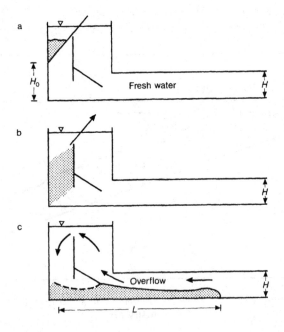

These experiments have led to some understanding of the transport and deposition of the sediments carried by turbidity currents. In one typical series the material used for the sediment was a moderately well sorted mixture of plastic beads of density 1.52 g cm^{-3} and mean diameter 0.18 mm. In these experiments the mode of deposition from the travelling bead suspension and the character of the vertical grading of the deposit were studied.

Study of the processes of erosion and deposition gave a picture as shown in figure 16.4, in which for a short distance behind the head the erosion process dominates, followed by an area of increasing deposition.

16.1.2 High-density turbidity currents

Details of the internal flow of a suspension current as shown in figure 16.4 are usually not visible from outside, but have been deduced in field work by the use of pressure sensors on the ground beneath the flow. Sharp separation of the fluid into two (or more) regimes has been established.

Laboratory experiments on high-density turbidity currents have been used to examine such two-phase suspension currents. Figure 16.5 shows one such experiment in progress (Taira, 1985), with the low-density Newtonian flow above, and the high-density 'body' of non-Newtonian flow beneath it. The initial composition of the fluid used in this experiment was bentonite (8%), sand (38%) and water (54%), forming a slurry of relative density 1.4.

16.2 Air suspension currents: fluidisation

Some of the most violent gravity currents occurring in nature depend on suspension of solid material by air or other gases. Examples are avalanches of airborne snow and pyroclastic flow eruptions from volcanoes, both of which may travel at speeds of over a 100 miles per hour.

16.2.1 Fluidisation experiments

An important process in suspension gravity currents is 'fluidisation', in which the introduction of streams of gas between solid particles can make them take

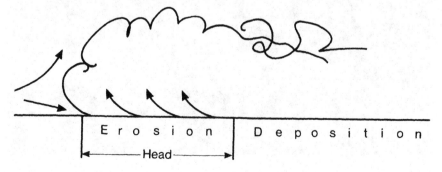

Figure 16.4 Areas of erosion and deposition at the front of a turbidity current.

on some of the properties of a fluid (Davidson & Harrison, 1963). For instance, fluidised material will flow down a slope, forming a kind of gravity current of particles suspended in air.

A favourite demonstration of the process is illustrated in figure 16.6 in which a toy plastic duck is placed at the bottom of a container with a layer of uniform dense particles. When a stream of air is passed through holes in the base of the tank, the particles are separated from their neighbours by the air rising between them and eventually each particle is supported by the drag forces due to the rising air. The contents of the tank begin to move around violently, looking like a boiling fluid; at this point the buoyant duck floats to the surface.

The apparatus used in some experiments (Wilson, 1980) with a gaseous fluidising medium is shown schematically in figure 16.7. A high pressure air supply is regulated and measured through a rotameter, then distributed into the fluidised bed. During runs, the height of the bed is recorded, together with the pressure drop on a water manometer. Using small smooth glass spheres one can

Figure 16.6 A plastic duck floating in a fluidised material.

Figure 16.5 Laboratory two-phase turbidity current. (Courtesy of A. Taira.)

derive a fluidisation plot (figure 16.8). The pressure gradient, $\Delta P/H$, increases linearly with gas velocity in accordance with Darcy's law for porous media. When the gas flow is supporting the weight of the bed, the bed is said to be incipiently fluidised. Any additional increase in the gas flow leads to the generation of gas bubbles.

There is commonly some degree of hysteresis evident, i.e. a difference

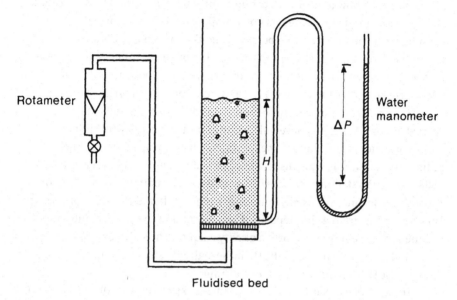

Figure 16.7 Schematic diagram of apparatus for fluidisation measurements. (After Wilson, 1980.)

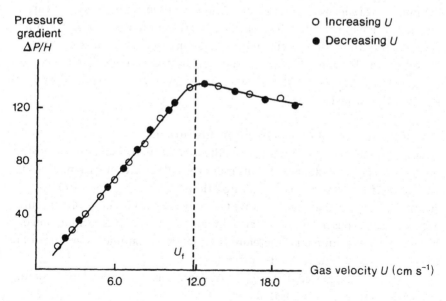

Figure 16.8 Variation of pressure gradient with gas velocity in fluidisation experiments. Incipient fluidisation occurs when $U = U_f$. (After Wilson, 1980.)

between measured curves for increasing and decreasing U. This is related to size distribution and to irregular particle shapes; the graph shown here, using uniform spheres, shows no hysteresis.

16.3 Initiation of turbidity currents

The problem of how a turbidity current is initiated, for example from a muddy slump on a slope at the sea bed, is not a simple one. The transformation is a complicated problem involving both soil and fluid mechanics. Slumping depends on the angle of slope, and the stability of the sediments; it may occur in earthquakes, even on slopes of only 2 or 3 degrees if shocks are sufficient (Morgenstern, 1967). Experiments in Lake Superior by Normark & Dickson (1976) have had some success in forming turbidity currents, but experiments carried out on the sea bed have failed to initiate any runaway turbidity currents.

Some successful laboratory experiments have involved setting the sediments in motion by means of a shock or vibration (Van der Knapp & Eijpe, 1968), simply by striking the base of the tank with a mallet. When a heap of homogeneous sand was used, such a shock merely altered the angle of its slope. However, using graded clay deposits it was found that the shock could cause the sediment mass to move and form a gravity current. The sediment mass started to move under the action of gravity, then the sliding system took up extra water and became sufficiently fluidised to continue.

An interesting example of an unplanned experiment on turbidity currents occurred in 1979 at Nice, on the Mediterranean (Gennesseaux et al., 1980). During the construction of an airport a 300-metre-long earthwork that was being built out into the sea slid away out of sight. About 400 million cubic metres of earth, and some earth-moving vehicles, vanished. Nearly four hours later, and 60 miles away, the resulting turbidity current cut the submarine telephone cable that joins Genoa and Majorca. Later, it cut the cable from Genoa to Sardinia. It was estimated that the speed of the current reached 25 miles per hour on the upper slopes of the sea bed.

16.3.1 Ignition of catastrophic turbidity currents

It is easy to imagine self-sustaining turbidity currents with a solid phase mostly of cohesive clay, since at low concentrations the fall velocity of such material is so low that it is only very slowly deposited. However, there is evidence on submarine canyons that highly erosive turbidity currents occur where the only available material is medium-to-coarse silts and sands. A proper formulation of bed-sediment entrainment is therefore essential for a clear understanding of an eroding and depositing turbidity current.

Parker (1982) showed that a necessary condition for a self-sustaining 'auto-suspension' current to exist is that

$$US/v > 1$$

where U is the mean down-channel flow velocity, S is the bottom slope and v is the sediment fall velocity.

It can be shown that a set of critical velocities and concentrations exists below which the turbidity current dies and above which it grows. Above the critical state the turbidity current is said to 'ignite', that is, it accelerates and entrains sediments, attaining the catastrophic state. Estimates of this catastrophic state suggest that it is highly erosive and capable of scouring out the submarine canyons which have been observed. This process of 'ignition' can take place in suspension flows both in water and in air, for example in the formation of avalanches of air-borne snow particles.

16.4 Summary of suspension processes

In figure 16.9 are sketched five different processes which can be involved in the maintenance of a suspension gravity current. Figure 16.9a shows a turbulent suspension of particles with low fall speeds, forming a gravity current whose behaviour is very similar to that of a saline gravity current. In figure 16.9b the particles have a larger fall-speed, and may bounce up from the ground, eroding further particles in the process. Both (a) and (b) can be expected in both liquids and gases.

The remaining three processes are all examples in which fluidisation plays a part. In figure 16.9c the fluidising material (either liquid or gas) enters the head of the gravity current from the surroundings. The ambient fluid is entrained in a series of Kelvin–Helmholtz billows above the head, but there is also a comparable flow of light fluid ingested beneath the foremost point of the nose. This can provide a supply of buoyant fluid to fluidise the interior of the head.

In the two other fluidisation processes, which have been dealt with in more detail in Chapter 10, the fluidising material involved (a gas) is supplied in different ways. In figure 16.9d red-hot material from a volcanic eruption releases water vapour from the surface over which it moves, by heating surface water or ground water, or fluids contained in plants. This fluidisation process appears to be an important one in nature. In the final sketch, figure 16.9e, the fluidising gas is produced from the solid particles, either by fracture or by diffusion.

In any particular example of a suspension current in nature, it is likely that more than one of these processes will be operating.

Bibliography

Allen, J.R.L. (1971). Mixing at turbidity current heads, and its geological implications. *J. Sediment. Petrol.*, **41**, 97–113.

Davidson, J.F. & Harrison, D. (1963). *Fluidised Particles*. Cambridge: Cambridge University Press. 155 pp.

Gennesseaux, M., Mauffret, A. & Pautot, G. (1980). Les glissements sous-marins de la pente continentale nicoise et la rupture de cables en mer Ligure. *C. R. Acad. Sci. Paris*, **290**, (Ser D), 959–62.

Figure 16.9 Summary of processes in suspension currents.

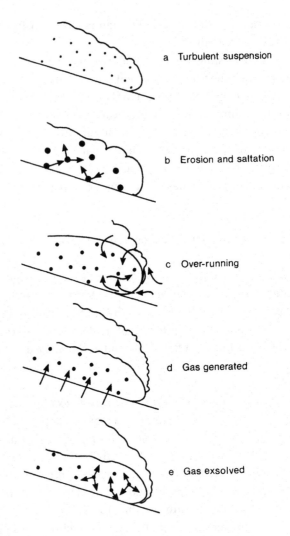

a Turbulent suspension

b Erosion and saltation

c Over-running

d Gas generated

e Gas exsolved

Kuenen, Ph.H. (1951). Turbidity currents of high density. *18th Int. Geol. Congr. London, Reports*, Part 8, p.44.

Middleton, G.V. (1966). Experiments on density and turbidity currents. I. Motion of the head. *Can. J. Earth Sci.*, **3**, 523–46.

Morgenstern, N.R. (1967). Submarine slumping and the initiation of turbidity currents. In *Marine Geotechnique*, ed. A.F. Richards, pp. 189–220. Urbans: University of Illinois Press.

Normark, W.R. & Dickson, F.H. (1976). Man-made turbidity currents in Lake Superior. *Sedimentology*, **27**, 815–31.

Parker, G. (1982). Conditions for the ignition of catastrophically erosive turbidity currents. *Mar. Geol.*, **46**, 307–27.

Riddell, J.F. (1969). A laboratory study of suspension-effect density currents. *Can. J. Earth Sci.*, **6**, 231–46.

Taira, A. (1985). Rheology and flow processes of sediment gravity flows. *Earth Monthly*, **1**, 391–7. (In Japanese.)

Wilson, C.J.N. (1980). The role of fluidisation in the emplacment of pyroclastic flows: an experimental approach. *J. Volcano. Geotherm. Res.*, **8**, 231–49.

Van der Knapp, W. & Eijpe, R. (1968). Some experiments on the genesis of turbidity currents. *Sedimentology*, **11**, 115–24.

17   Gravity currents on a rotating earth

Which way does the bath-water rotate as it runs out? Many people have been intrigued by questions about the effect of the earth's rotation on outflowing bath-water and have looked to see whether its sense of rotation is opposite in the northern and southern hemispheres.

In the bath-water example and in many gravity currents the effect of the earth's rotation is very small indeed and can be neglected. Nevertheless, in many large-scale environmental gravity currents the earth's rotation exerts an important influence; one example already mentioned in Chapter 5 is the flow in the Southerly Buster.

17.1 Coriolis force

Because all our measurements are made on a rotating earth, the path of an object which a stationary observer in space sees as a straight line will appear curved to us. We can illustrate this by the simple experiment shown in figure 17.1, where a ball is rolled in a straight line from the centre towards the edge of

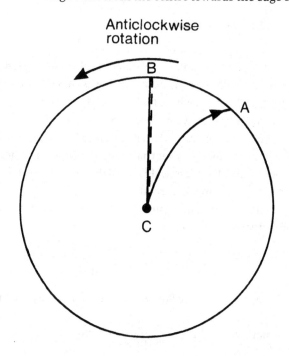

Figure 17.1 Demonstration of Coriolis force, using a rotating disc. A ball is rolled from C towards B, but in a frame of reference moving with the disc it is deflected towards A.

Anticlockwise rotation

a disc rotating in an anticlockwise sense. As expected from Newton's First Law of Motion, the ball continues in uniform motion in a straight line relative to a fixed frame of reference (for example the room containing the rotating disc), but viewed relative to coordinates moving with the disc, the ball swings to the right of its initial line of motion. In the analogous case of the earth, using our usual rotating reference frame of latitude and longitude, there is a deflection of moving objects to the right in the northern hemisphere and to the left in the southern hemisphere.

So, living as we do in a rotating system, when we attempt to move something in a straight line it is deflected sideways. The apparent force, required by Newton's Second Law of Motion to produce this change of momentum in the rotating frame, is called the Coriolis force.

The Coriolis force acting on a body is proportional to the speed of the body and the angular velocity of the earth (close to 2π radians in 24 hours). The force, per unit mass, also depends on the latitude and can be written as

$$\text{Coriolis force} = fV$$

where $f = 2\Omega \sin \phi$, and is known as the Coriolis parameter, Ω is the angular velocity of the earth (7.29×10^{-5} radians s^{-1}), V is the velocity of the moving object, and ϕ is the latitude. Near the equator, $f \approx 0$, and rotational forces are less important; for example, there are no equatorial cyclones.

17.2 Rossby number

The dimensionless number which describes the relative importance of inertial and rotational forces in a rotating flow is called the Rossby number; this can be written as Ro$=V/fL$, where f is the Coriolis parameter and V and L are characteristic velocity and length scales of the flow.

The effect of the earth's rotation can be neglected in rapid small-scale flows with Ro very much greater than 1, but becomes significant when the value of Ro is about unity. For large-scale flows which evolve over days, Ro <1 and the effects of rotation become dominant.

The Rossby number may be described as the ratio of the rotational and convective time-scales

$$\text{Ro} = \frac{1/f}{L/V} \quad \text{which is equivalent to} \quad \frac{1 \text{ day}/2 \sin \varphi}{\text{time-scale of flow}}$$

The approximate Rossby numbers for three previously mentioned flows are:

$$\text{Bath water: Ro} = \frac{1 \text{ day}/2\sqrt{2}}{1 \text{ minute}} \approx 1000$$

Sea breeze: $\text{Ro} = \dfrac{1 \text{ day}/2\sqrt{2}}{\frac{1}{2} \text{ day}} \approx 1.4$

Southerly buster: $\text{Ro} = \dfrac{1 \text{ day}/2\sqrt{2}}{3 \text{ days}} \approx 0.2$

The structure and dynamics of gravity currents in rotating systems are found to be different from those without background rotation. In the absence of boundaries, a gravity current released from a point and moving radially in a rotating frame will approach a state of equilibrium in which no further spread is possible. With vertical or inclined boundaries present, for example in a rotating lock exchange system, the Coriolis force generally makes the flow hug the boundaries and so leads to three-dimensional boundary currents.

17.3 Spreading under gravity: no boundaries

When a fluid spreads under gravity in a rotating system, in the absence of boundaries, instability and viscous dissipation, the flow approaches a state of equilibrium. This state is reached when buoyancy and Coriolis forces are in balance, and further release of potential energy is impossible.

Most of the large-scale flows that we see in weather maps have reached the stage where they are dominated by the earth's rotation. For example, the wind does not blow straight down the density gradient into the centre of low pressure, but has turned to the right and blows around the centre of low pressure.

Gravity-driven flows have been examined in a water tank mounted on a rotating table. Various types of instability have been observed at the edge of the dense flow after outward movement has ceased, and figure 17.2 shows an

Figure 17.2 View from above of instability at the edge of a gravity current spreading in a rotating flow. (Courtesy of P.F. Linden.)

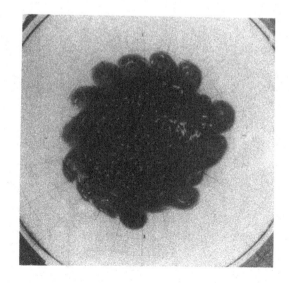

example in which a series of large eddies can be seen, rolling up along the leading edge (Griffiths & Linden, 1982).

17.4 Spread along a boundary

The behaviour of a gravity current produced by lock exchange in a rotating laboratory tank is illustrated in figure 17.3. In a rotating system, within a rotational time-scale $1/f$, the Coriolis forces become comparable to the buoyancy forces and there is no motion in the direction along the current except along the walls. The 'head' and the trailing flow exist much as in the non-rotational gravity current, and the mixing and billows are of the order of the current width; however, there are also rotation-dominated instabilities which have much longer length scales.

The nose velocity, depth and width all decrease with time. The width scales on $(gH)^{\frac{1}{2}}/f$, where H is the fluid depth. This depth scale is referred to as the Rossby radius of deformation.

The nose velocity at the wall scaled by local values of $(g'H)^{\frac{1}{2}}$ is constant, as expected from non-rotational theory, and the flow is independent of Reynolds number, VH/ν, for values of Re > 1000. Figure 17.4 shows the flow of a buoyant current along the right-hand wall after release from a lock in a laboratory channel with anti-clockwise rotation (Griffiths & Hopfinger, 1983).

Bibliography

Griffiths, R.W. & Hopfinger, E.J. (1983). Gravity currents moving along a lateral boundary in a rotating fluid. *J. Fluid Mech.*, **134**, 357–99.

Griffiths, R.W. & Linden, P.F. (1982). Laboratory experiments on fronts. *Geophys. Astrophys. Fluid Dynam.*, **19**, 159–87.

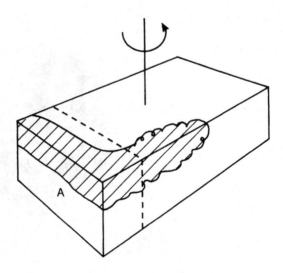

Figure 17.3 The spread of a buoyant gravity current after release from a lock at the end 'A' of a rotating laboratory channel.

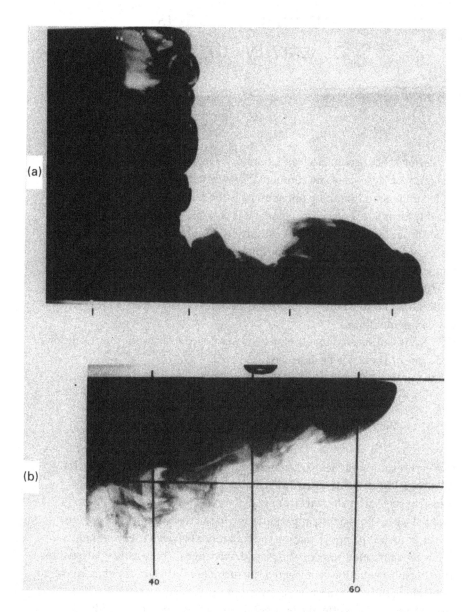

Figure 17.4 (a) Top view and (b) side view of a buoyant gravity current after release in a rotating channel in the conditions shown in figure 17.3. (Courtesy of R.W. Griffiths.)

18 Numerical models of gravity currents

As previous chapters make clear, laboratory experiments can elucidate the dynamics of large-scale environmental gravity currents. Another powerful tool to understanding these phenomena is the numerical model. Using a suitably formulated numerical model it is possible to examine separately a whole range of parameters. For example, in thunderstorm outflows the model can examine the full range of atmospheric parameters (such as wind, temperature stratification and moisture) which might affect outflow behaviour. In addition, the model can be used to reproduce the basic results of laboratory studies in order to identify any differences between laboratory and environmental flows.

Several different models have been developed to examine dense outflows, and some of them will be described.

18.1 Marker-and-cell technique

One early example is the marker-and-cell numerical technique used by Daly & Pracht (1968) to study a two-dimensional gravity current surge in the transient stage after the start from rest. A solute transport equation was coupled with a Boussinesq approximation to the Navier–Stokes equations. (In the Boussinesq approximation the difference between the two fluids is neglected in momentum equations, and only appears in density relationships.) The transport equation represented the turbulent diffusion which is known to occur along the shear interface between the gravity current and the environment. In this effectively one-fluid model, the use of this eddy diffusivity in the solute transport equation represents the mixing quite satisfactorily.

Figure 18.1 shows a series of marker particle plots in the development of a simulated gravity current. The grid mesh is 60 by 20 points, and the ratio of the densities is 1.2. The plots on the left are from a full two-fluid model, and the right series uses the Boussinesq approximation in a 'one-fluid' solute transport model. The results are quite similar, except for a slightly blunter nose in the one-fluid simulation.

A comparison of these results with measurements of laboratory gravity currents has been made and the results are in good agreement.

18.2 Models of thunderstorm outflows

Some numerical models were specifically designed during the 1970s to examine the flow of the cold dense air which descends from a thunderstorm and spreads out horizontally as a gravity current.

One such model (Mitchell & Hovermale, 1977) was formed with a time-dependent heat sink designed to represent the evaporative cooling associated with thunderstorm downdraughts. The lateral boundaries were rigid walls, so no environmental flow could be incorporated, and surface friction was introduced through a drag term in the lowest grid level. Figure 18.2 shows a vertical cross-section of the potential temperature field 12 minutes after the start. The cold air outflow is advancing from left to right, away from the parameterised downdraught, and a clear head structure is evident.

The limitations of this work, for example rigid boundaries and a small computational domain, were largely a consequence of the limited computer resources available at the time. Their study dealt with many questions concerning outflow dynamics, for example the sensitivity to ambient stability,

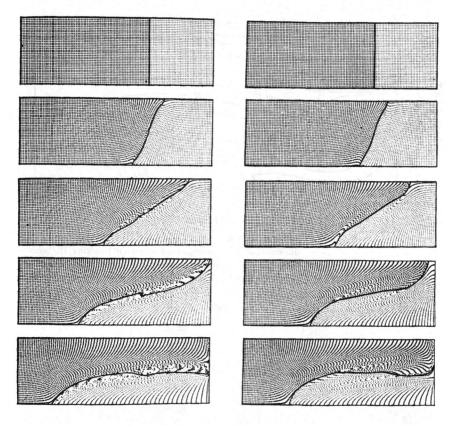

Figure 18.1 A two-dimensional gravity current surge modelled by marker particle plots. The left series was produced by a full two-fluid model, and the right series by a Boussinesq approximation to two-fluid flow. (From Daly & Pracht, 1968.)

surface friction, and the density difference across the gust front, and laid the foundation for future outflow modelling studies.

18.2.1 A three-dimensional outflow model

At about the same time as the work described above, an attempt was made by Teske & Lewellen (1977) to model the geometry of outflows with an axisymmetric outflow model. This also included modelling of the turbulence structure of the flow. Being axisymmetric, ambient wind-shear could not be included in the model. Figure 18.3 shows the development of the temperature field of the outflow, illustrating clearly the development of the outflow and its associated head as the downdraught strikes the ground. There is a wavy structure behind the head along the top of the cold air current moving into the domain initially at rest.

18.3 Analytical and numerical model

A 1980 model by Thorpe *et al.* presented analytical and numerical results of two-dimensional updraught and downdraught flows with emphasis on

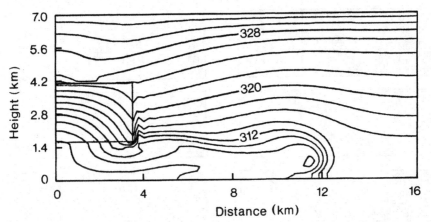

Figure 18.2 Potential temperature (in K) shown in a two-dimensional model, using a time-dependent heat sink. The box shows the position of the parameterised downdraught. (From Mitchell & Hovermale, 1977.)

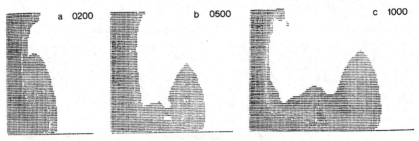

Figure 18.3 An axisymmetric outflow simulation, showing a cross-section of the development of the temperature field. (From Tesker & Lewellen, 1977.)

thunderstorm outflows. This employed observed downdraft profiles, and results included the generation of convection along advancing gust fronts. It was found that the basic flow features were described quite accurately.

This numerical model to examine outflow dynamics had a time-dependent, dry, two-dimensional form. The types of updraughts and downdraughts used in their model are shown schematically in Figure 18.4. The downdraught was introduced at the right-hand boundary, and the outflow proceeded across the domain from right to left. The domain was divided into five distinct regions.

Region I Steady downdraught flow with large horizontal gradients.

Region II Steady horizontal flow with small horizontal gradients.

Region III The gust front region.

Region IV The relatively calm ambient environment ahead of the front.

Region V The calm environment well above the outflow.

The model domain had dimensions 80 km horizontal distance by 720 mb vertical pressure range, with grid spacing 500 m by 40 mb, respectively; however, the model lateral boundaries allowed gravity waves and the gravity current head to pass out of the domain.

A simple eddy viscosity was incorporated, with no surface friction. The model was run with a uniform ambient wind, but with no vertical wind-shear. The time-step of the numerical scheme was 15 seconds.

The main drawback to the results of this work was the poor resolution of the numerical model. Observations and laboratory experiments have shown that gravity currents are highly turbulent phenomena, and the numerical model could only represent the bulk flow of such a current. It appeared that higher spatial resolution was needed if numerical models were to resolve the turbulent structure of gravity currents.

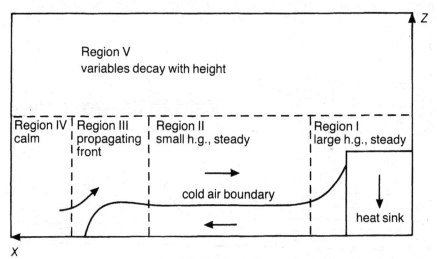

Figure 18.4 A schematic diagram showing the various regions of flow indentified in the outflow modelling by Thorpe *et al.*, (1980). h.g.=horizontal gradient.

18.3.1 Gravity currents and bores

Section 13.1.2 described the production of undular bores in the laboratory using a saline gravity current in a two-layer tank. In a two-dimensional model to study this phenomenon (Crook & Miller, 1985), a gravity current was initiated at the left boundary, and while it was moving across the domain from left to right, a low-level stable layer was created in the right half of the domain by applying a cooling function near the ground. When the gravity current interacted with the dense layer, it triggered an undular bore which propagated across the domain, as shown in figure 18.5.

18.4 High-resolution models

Laboratory experiments show that gravity currents usually contain shearing instabilities, but none of the numerical models described so far can reproduce them. The grid resolution is too coarse to resolve the physical turbulence known to be present at gravity current heads.

A two-dimensional numerical model capable of investigating billows at a gravity current front has been developed by Droegemeier & Wilhelmson (1985). The philosophy of this model was to avoid making any approximations to the governing equations. Instead of parameterising turbulence, very high spatial resolution was used to resolve the important small-scale features. A weak background smoothing term was applied to discourage the growth of spurious computational instabilities. The lower boundary was a rigid, free-slip plate.

The model domain was 30 km long and 10 km high with a grid spacing of 100 m in each direction. The flow was initiated as a purely horizontal flow by placing a 2 km high column of cold fluid, with density 2% greater than the

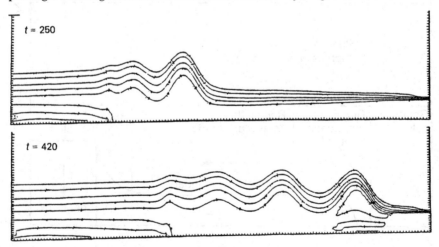

Figure 18.5 Potential temperature contours of a numerically simulated undular internal bore. (From Crook & Miller, 1985.)

environment, at the left boundary, and this was held fixed throughout the simulation.

Initially the outflow was a laminar current with a speed of 18 m s^{-1}. By 10 minutes, wave-like perturbations had developed in the region of sharp density contrast at the top of the outflow. In time these perturbations rolled up into horizontal vortices characteristic of Kelvin–Helmholtz billows. Figure 18.6a shows the density field after 20 minutes, using the 100 m grid spacing. To illustrate the effect of changing the grid spacing, figure 18.6b shows an identical simulation, but with a 200 m grid spacing, and figure 18.6c shows an identical simulation with a 500 m spacing. It is clear that a grid spacing as fine as 100 m is necessary to model the development of the billows, which have characteristics close to those described in Chapter 11.

The reasons why earlier models did not show evidence of Kelvin–Helmholtz instability include both the lack of spatial resolution and the use of too much

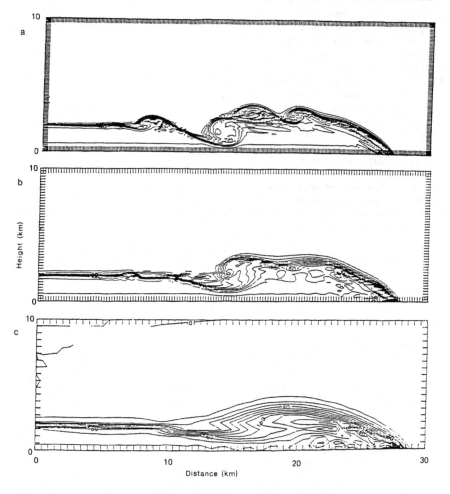

Figure 18.6 Three density contour plots of the simulation of a gravity current head. Grid spacings are (a) 100 m, (b) 200 m, (c) 300 m. (From Droegemeier & Wilhelmson, 1985.)

numerical smoothing. It seems that the instability along the top of the outflow is very sensitive to the magnitude of the computational mixing.

The time of writing is one of rapid development of numerical models, and we can look forward to fine-resolution models in three dimensions. These will study the full development of Kelvin–Helmholtz billows and perhaps also the instability shown in the lobes and clefts.

Bibliography

Crook, N.A. & Miller, M. J. (1985). A numerical and analytical study of atmosphere undular bores. *Q. J. R. Meteorol. Soc.*, **111**, 225–42.

Daly, B.J. & Pracht, W.E. (1968). Numerical study of density current surges. *Phys. Fluids*, **11**, 15–30.

Droegemeier, K.K. & Wilhelmson, R.B. (1985). Kelvin–Helmholtz instability in a numerically simulated thunderstorm outflow. *14th Conf. Severe Local Storms, Indianapolis*, pp. 151–4. Boston: American Meteorological Society.

Mitchell, K.E. & Hovermale, J.B. (1977). A numerical investigation of a severe thunderstorm gust front. *Mon. Wea. Rev.*, **105**, 657–75.

Teske, M.E. & Lewellen, W.S. (1977). Turbulent transport model of a thunderstorm gust front. *10th Conf. Severe Local Storms, Omaha*, pp. 143–9.

Thorpe, A. J., Miller, M.J. & Moncrieff, M.W. (1980). Dynamical models of two-dimensional downdraughts. *Q. J. R. Meteorol. Soc.*, **106**, 463–84.

Index

Aberfan, mud flow 1966, 119
acoustic fluidisation, 118
acoustic sounding, 18, 64
aegre, Trent bore, 96
aerosol
    Lake Nyos, 136
    sea-breeze front, Japan, 40
air avalanche, katabatic gust front, 64
air cavities in ducts, 171
air fluidisation, 118
aircraft hazards, 2, 66-9, 173
    front of gravity current, 66
    microbursts, 67
    take-off and landing, 66
    warnings, 69
    wind shear, 69
ambient stratification, 186
    two-layer system, 186
    uniform density gradient, 200
andhi (thunderstorm outflow), 14
anvil cloud, 12, 53
aphids, at sea-breeze fronts, 40
Araguari River, bore, 96
arc clouds, 46
Armero, mud flow, 134
armyworm moth, 42
arrested saline wedge, 145, 154
atmospheric bores, 21-3
atmospheric gravity currents, 11-27
atrium, 83-5
auto-suspension currents, 224
avalanches, 112-22
    airborne snow, 2, 224
    fluidisation, 114
    mud, 118-22
    rock, 116-18
axisymmetrical collapse
    experiments, 172-7
    numerical model, 234

barrier to gravity current
    sloping above, 160
    porous, 158
    solid, 159
basal melting theory, fluidisation, 118
basaltic lava, 127
    levees, 127
    Surtsey, 127
    tunnel, 127

bath water, rotation in outflow, 227-8
Bernoulli's equation, 143-4
Bhopal, 1984, dense gas disaster, 70
billows at front, 142
    Kelvin-Helmholtz, 147, 149, 225, 237
    numerical modelling, 238
birds, at sea-breeze fronts, 40
black smokers, 137
blocking, in stratified fluids, 187
    complete, 192
boom, to contain oil slicks, 108
bores
    axisymmetric, 196
    energy loss, 7
    generation of, 24, 35, 186
    internal, 7, 99-101
    Loch Ness, 99
    tidal, in rivers, 5, 97-9
    two-dimensional numerical model, 236
    undular, 5
    wavelength, 25
Boulder, Colorado, cold air drainage, 64
Boussinesq approximation, 181, 232
box models, 71, 169, 172
Bradford football ground fire, 86
Brunt-Vaisala frequency, 198
bubbles, rising, for turbulence, 208
buoyancy
    flux, 175
    frequency, 198
    inertial regime, 214
    viscous regime, 214
Burketown, N. Australia, 23
bursting process in mudflows, 121

cables, submarine, cut by turbidity currents, 102,
    103, 224
Cambrian colliery, 1965 explosion investigation,
    78
canyons, submarine, 102
catastrophic ignition in turbidity currents, 224
cavity flow, 143
Celtic Sea front, 91
circular ducts, lock exchange flow, 166
Cleveland, 1944 explosion, 69
CO_2 saturation, Lake Nyos, 136
collision
    gravity currents, 196
    sea-breeze fronts, 23, 29-34

cold air drainage, 64
cold fronts, 49–52,
Connecticut River front, foam line, 92
constant flux
   parallel channel, 175
   radial flow, 177
cooling-water requirements, 105
Coriolis
   bath water, 227–8
   force, 59, 227–30
   Rossby number, 228
   sea breeze, 229
   southern hemisphere, 59
   Southerly Buster, 229
Cousteau Society, 96
Crete, Nebraska, 1969 gas explosion, 70
crossing fronts, 197
curtain clouds, 34
custard powder, suspension flow, 121

dam-break, 3
dam-break analogy, 4, 167
de Saussure, on debris flows, 118
debris and mud avalanches, 112, 118–22
decreasing depth, flow into, 160
demolished building, suspension current,
   218
dense gas, disasters
   Bhopal, 1984, 70
   Montanasm, 1981, 70
   Cleveden, 1944, 69
   Crete, Nebraska, 1969, 70
   Los Alfraques, 1978, 70
   Meldrim, 1959, 70
   Mexico City, 1984, 70
   Potcherstroom, 1973, 70
dense gas dispersion, 2, 69–76
   models, 70–2
   field experiments, 72–6
density jump, 177
diffusion, turbulent, 208
diffusivity, eddy, 210
door, gravity current through, 2, 81
doppler radar, 19
downwind fronts, 73
drainage of cold air, 64
duck, plastic fluidised, 222
dust
   airborne, 14
   cloud, 1
dynamic similarity, 11

Elm, rockfall, 1881, 116
energy loss, at bore, 7
erosion
   of channel by turbidity currents, 219
   by turbulence, 207
estuaries, fresh water gravity currents, 2

fires in buildings
   Bradford football ground, 86
   King's Cross , 86
   spread along ceiling, 85
Fisher River, salt wedge, 92
fjords
   fronts in, 2, 94
   Loch Eil, 95
   Loch Etiven 95
   waves in stratified layer, 200
floating debris at front, 88–90, 92
flow force, 143
flow on rotating earth, 59, 227–31
fluidisation, 221–4
foam lines, 2, 92, 101
foremost point of gravity current, or nose, 140,
   152
Fraser River, Canada, 2
free surface flow, 165, 178
frontogenesis, 32
fronts, anatomy of, 140–63
fronts, in the environment
   fjords, 94
   estuaries, 91
   ocean, 88
   sea-breeze, 2, 29–43, 211
   thunderstorm outflows, 1, 12–18, 233–4
fronts, formation and dissipation, 206, 210
Froude number, 143, 147, 149, 191, 200
Fujita, Professor T.T., 68
Fundy, Bay of, 99

gas, spread of dense, 69–76
Gibraltar Strait, flow over sill, 100
glacier
   erosion, 125
   galloping, 124
   speeds. 123
   transportation of material, 125
   winds, 63
gliders
   curtain clouds, 34
   flights of 9 June 1968, 30
   Lasham Gliding Centre, 31, 33, 35
   sea-breeze fronts, 29, 31
   slope upwinds, 63
   swifts, flight with, 40
Golborne Colliery, 1979 accident, 79
Grand Banks, sediment slide, 102
gravity waves, 9
grid
   oscillating, 206
   turbulence, 206
Gulf Stream, 88
gust front detection, 69

H_2 S in Lake Nyos, 136
haboob, sand and dust storm
   lobe and cleft structure, 14

raised nose, 140
thunderstorm outflow, 14
Hangzhou, Qiantang bore, 96
head, of gravity current, 140
    cut off in ambient stratification, 201
    form on slope, 179
    head and tail winds, 153–4
Health and Safety Executive
    mine accidents, 78
    field experiments, 73–4
heat sink, in numerical model, 233
HEGADAS, dense gas model, 75
house, ventilation, 2, 81–2
hydraulic jump, 5, 155, 169, 177, 179
hydrothermal vents
    black smokers, 137
    mega-plumes, 138
hysteresis, in fluidisation experiments, 223

Ice Ages, 125
ignition, of suspension current on slope, 224
immiscible fluids, laboratory experiments, 108
inclined thermal, 180
inertial regime, 214
insects, at sea-breeze fronts, 41
instablilities
    billows, 142, 147, 149, 225, 237
    laboratory suspension flows, 121
    lobes and clefts, 14, 19, 34, 62, 142, 147, 149, 238
    viscous flow on slope, 215
internal bores
    atmosphere 7
    generation by gravity current, 101, 189–92
    lakes, 99
    ocean, 8, 99–100
internal solitary wave, 10
internal waves, in continuous stratification, 200
intrusions, 194, 203
inversion, nocturnal, 21
inviscid-flow theory, 142–4
Ionica building, 85

Jiang-jia Ravine, mud flows, 121
jökulhlaup, 134

katabatic winds
    air avalanche, 64
    fronts, 63–4
    glacier winds, 64
    Mawson, Antarctica, 64
    Red Butte Canyon, 64
Kelvin–Helmholtz billows, 147, 149, 223, 237
King's Cross fire, 86
Knight Inlet, flow over sill, 100

LNG, liquid natural gas, 69–70, 75
LPG, Liquid propane gas, 75
lahar, or mud flow, 119, 133
Lake Nyos, Cameroon, gas disaster, 136

Lake Superior, turbidity currents, 224
lakes, plunge line, 95
landslides, submarine, 102
laser 'radar', *see* lidar
Lasham Gliding Centre, 31, 33, 35
lava, 127–8
layering number, in mines, 78
lid, rigid sloping, 160
lidar, 20
lobes and clefts, at front, 14, 19, 34, 62, 142, 147, 149, 238
Loch Eil, renewal of deep water, 95
Loch Etive, 95
Loch Ness, internal bore, 99
lock exchange flows, 164
locusts
    at sea-breeze fronts, 42
    outlining front, 43
Los Alfraques, 1978 disaster, 70
low Reynolds number experiments, 213

mamma cloud, at thunderstorm, 13
Maplin Sands, field experiment, 75
Marco Polo, at Hangzhou, 98
marker and cell, numerical model, 232
Mawson, Antarctica, katabatic winds, 64
mechanical fluidisation, 118
Meldrim, gas explosion, 1959, 70
Mexico City, gas explosion, 1978 and 1984, 70
microbursts, 67–8
Middlesbrough, smog, 38
mines
    Cambrian Colliery, 78
    gases in, 76–80
    Golborne Colliery, 79
    layering number, 78
mixing in gravity currents, 144, 207
Moncton, New Brunswick, bore in Petitcodiac River, 99
Montanas, 1981 gas disaster, 70
Morning Glory, 9, 21–3, 47
motor-gliders, 32
Mount Augustine, Alaska, 133
Mount St Helens, 1980 eruption, 135–6
moving-floor experiments, 206
mud flows
    bursting process, 121
    instabilities, 120
    Jiang-jia Ravine, China, 121
    Mt St Helens, 135
    increase in frequency, 120
    speed of, 119
    Stava disaster, 1985, 120
    superelevation in canyon, 134
    volcanic eruptions, 133
    White Mountain, observations, 119

Nevado del Ruis, Columbia, lahars from eruption, 134

New Brunswick, Canada, radar records, 42
Ngauruhoe, New Zealand, nuée ardente, 131
Nice, France, 1979 unplanned turbidity current,
    220
non-Newtonian flow, 221
nose, foremost point of gravity current, 140,
    152
nuées ardentes, 127, 130
numerical models
    axisymmetrical outflow, 234
    dense gas spread, 70-2
    HEGADASII, 75
    gravity currents, 232
    high resolution, 235
    marker and cell, 232
    thunderstorm outflow, 233
    undular bore, 236

obstacles, to gravity currents, 155
oil slicks, 12, 106-11
    containment of, 108
    gravity–inertial regime, 108
    gravity–viscous regime, 108
    laboratory experiments, 108
    time-scales, 108
    viscous–surface regime, 108
oscillations, constant flux radial flows, 178
overflow, 178

pampero secco, 62
Pelee, Mt Martinique, pyroclastic flows, 129
Petitcodiac River, tidal bore, 99
plunge point, lakes and reservoirs, 95
pollution
    in fjords, 95
    in sea breeze, 37
Potcherstroom, 1973 disaster, 70
porous
    barrier, 156
    floor, 182
Porton Down, field experiments, 72
potential temperature, 11
power station effluents, 105
precipitation roll, at thunderstorm outflow,
    20
pressure changes, surface, 15, 21
pressure jump lines, 16
processes in maintaining suspension, 225
pulsing flow of gravity current, 178, 201
purging through duct, 162
pyroclasts
    into water, 133
    Mt St Helens, 135
    Mt Ngaurohoe, N131
    pyroclastic gravity current, 129-33
    pyroclastic plumes, 128

Qiantang River, tidal bore, 96-9
Quinault River, series of lines in flow, 101

radar
    sea-breeze front, 41
    thunderstorm outflows, 19-20
Red Butte Canyon, katabatic wind, 64
reduced gravity, 4
release of fixed quantity of fluid, rectangular
    channel, 166-72
remote sensing
    lidar, 20, 40
    radar, 19, 41
Reynolds number
    definition, 11
    effect on profile, 140
    effect on height of nose, 152
    flow under sloping lid, 161
    ratio of inertial to viscous forces, 214
    limits of dependence, 11, 220
Richardson number, 77-8, 147, 154
ring vortex, at microburst, 68
Rio Magdalena, Columbia, turbidity currents,
    103
rivers
    estuarine fronts, 91-3
    salt wedges, 91-2, 145, 154
    tidal bores, 5, 97-9
rock avalanches, 116-18
    Mt St Helens, 135
    fluidisation mechanisms, 118
    travel distance, 116
roll cloud, 9, 22, 59
rope clouds
    cold front 49
    hurricane, 56
    thunderstorm outflow, 46
    tropospheric, 53
    warm front. 54
Rossby number, 228
rotating earth, effects on gravity currents,
    227-31
Russell, Scott, 10

salt wedge, arrested, 91-2, 145, 154
Sargasso Sea, 88
satellites
    geostationary, 45
    polar orbiting, 45
satellite imagery
    mergers, 47
    Morning Glory, 47
    oceanic fronts, 88
    rope clouds, 46, 49, 53
Scott Russell, 10
sea breeze, 2, 23
sea-breeze fronts
    birds, 40
    bore development, 35
    Chicago, 37
    cloud forms, 34
    collisions, 23

forecasting, 30
frontogenesis, 31, 211
gliding flights at, 30
insects at, 40
Lasham Gliding Centre, 31, 33, 35
lidar sensing, 40
pollution, distribution of 37
progress inland, 31, 38
radar, 41
Riverside,California, 37
structure of, 30
Tokyo, 40
vortex, formed in evening, 35
sector, flow in a 172
sedimentary rocks, 219
Seiont River, Caernarfon, 92
self-similar solution, release of fluid, 169
self-stoking, suspension currents, 5
Seneca Lake, internal bore, 100
Severn River, bore, 5, 96
Seville University building, 84
shadowgraph technique, 146
shallow sea front, 88
shock-initiation of turbidity current, 224
silicone oils, for viscosity experiments, 214, 216
sill, flow over
  Knight Inlet, 100
  Gibraltar Straits, 100
similarity, dynamic, 11
Sizewell power station, effluent, 106
slopes
  gravity currents, 178
  viscous flows, 215
sloping lid, flow beneath, 160
smog, 37
snow avalanches
  airborne powder snow, 113–14
  barriers, 116
  flow avalanches, 113
  giant snowballs, 112
  runout distance, 116
  starting zone, 113
  triggering, 113
solid barrier, 155
solitary waves, 9, 21, 155, 197
sonar, acoustic radar, 18
Souffriere Volcano, St Vincent, extruded lava,
  128
Southerly Buster, 59–62
speeds
  bores generated by gravity currents, 191
  gravity currents, 3, 17
  mud flows, 134
  thunderstorm outflows, 17, 19, 26
spruce armyworm, 42
squalls from thunderstorms, 1
stagnation point, 147
starting plumes, 176
Stava, mudflow from dam-burst, 120

stratification, ambient, 186–204
structure of thunderstorm outflow, 14
Sturzstrom, 116
subcritical flow, 186, 189, 191
superelevation, of mudflow in canyon, 137
supercritical flow, 186, 189, 191
surface-active materials, 178
surface tension, 172, 178, 214
surfers, on bore, 6
Surtsey, Iceland, lava flow, 127
suspension flows
  kaolin, 5
  turbidity currents, 219–25
  two-phase, 221
swifts, at sea-breeze fronts, 40

Tehuantepecer, 48
temperature difference, effect of, 4
termite mound ventilation, 82
tetroons, 37
thermal, inclined, 180
thixotropic material, 121
Thorney Island, dense gas field experiments, 72
thunderstorm outflows
  airborne dust, 14
  acoustic sounding, 18
  cell structure, 13
  depth of flow, 17
  ground observations, 13
  gust front, 19
  numerical models, 233–4
  precipitation roll, 20
  pressure changes, surface, 15
  radar, 19
  speed of front, 17, 19, 26
  structure of front, 14
  squalls, 1
tidal bores, in rivers
  Araguari, 96
  Petitcodiac, 99
  Qiantang, 96, 99
  Severn, 5, 96
  Trent, 96
tilting tank experiment, 211
Tokyo, sea breeze, 40
Toute River, Mt St Helens, 136
tower measurements, 17, 24
transition phase, current down slopes, 180
trapezoidal lock exchange flow, 165
triangular lock exchange flow, 166
tropopause, 12, 53
tunnels
  lava, 127
  vehicular, 81
turbidity currents
  definition, 5
  erosion of channel, 221
  experiments in laboratory, 219
  high density, 221

turbidity currents (cont.)
  initiation of, 224
  lake, entering, 212
  ocean bed, 102
  two-phase, 221
turbulence and gravity currents, 205–10
Trent River, bore, 96
Trondheim Fjord, foam lines, 101
two-phase suspension currents, 221
two-tank method, for uniform stratification,
    197

underflow, 178
underground garages, 81
undular bores, 5, 186, 197
upwind fronts, 153

vehicular tunnels, 81
ventilation, flow through doors, 81
viscous gravity currents, 3, 140, 213–17
volcanic eruptions, 127
vortex ring at leading edge of front, 73, 173–4

warm fronts, 54
waves
  gravity, 9
  solitary, 9, 21, 155, 197
wedges, arrested, 91–2, 145, 154
White Mountain, California, mud flow
    description, 119
widening channel experiment, 174

Yangtze River, 96

Printed in the United States
By Bookmasters